WEBMASTERING TH

Webmastering The Craft

Fraternity in the Digital World
2020 Edition

Brother Ken JP Stuczynski

cyphrGlyffe

An Imprint of
Amorphous Publishing Guild, Buffalo, NY USA
Amorphous.Press/cyphrGlyffe
©2020

Webmastering [noun] – The work of aiding, organizing,
or leading a group's communications and
public relations in a modern way; serving by use of
skills and knowledge of the digital arts.

DISCLAIMER: This book contains guidance for webmasters and
others who serve the craft in digital media. Much of it is oriented
specifically for the jurisdiction of the Grand Lodge of the State of
New York. In general, these are meant to be explanatory rather than
authoritative, and if in doubt how to use this information, consult
your Grand Lodge through appropriate channels.

The author is publishing this work independently of his affiliation
with The Grand Lodge of the State of New York or any other
Masonic jurisdiction or body, and therefore no endorsement is
implied.

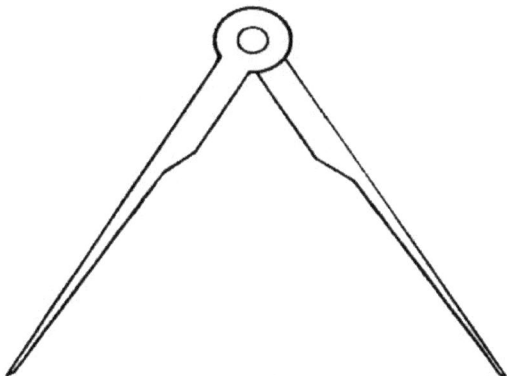

To all those who serve the Craft with their digital talents and skills.

CONTENTS

SECTION III. PUBLISHED ARTICLES

INTRODUCTION

This is not the first Masonic book I had hoped to publish as a sole author. My interest in the Fraternity is both personal and esoteric, and I've written at least a word or two on nearly all aspects of those. However, as someone often referred to as the "Grand Webmaster" in my jurisdiction, there is a greater Trestle Board I must heed. As I write this, most of the nation is sequestered due to the novel coronavirus, COVID-19.

Never in my lifetime have I seen our society squarely in need of digital technology to do everyday things. It is more painful to me that we could have made such processes and behaviors commonplace years ago. In this respect, we have even fallen behind nations that a few short years ago were without ubiquitous phone systems. Today, much of the world does their shopping, law, business, and finance online and by phone apps in ways that make us seem quaint.

Despite circumstances, much good is already coming of this. We are reaching out to check on Brothers and widows, perhaps with whom we should have kept better in touch previously. We are scrambling to hold meetings on video and audio conference platforms, and seeing Brothers we forgot were members because they couldn't make physical meetings. Some filling out of forms and other information exchanges are only done on websites and apps now. Discussions are becoming more prolific in email chains, texts, and even social media channels. And we may give much more consideration to the lawful use of digital signatures and seals without the need for emergency dispensations. Difficult times are forcing all of us to separate the wheat of what is really important from the chaff of superfluous bureaucratic tradition.

For over five years now, I have advocated for the use of technology in much of what we do as Freemasons. But I have equally advised against its use when a more human approach would be desirable. The chapters herein should not be meant as a textbook of strategic marketing or mass communication. We must always remember that the message must define the medium, not the other way around.

It is my hope that this work is not just timely, but will serve as a useful guide well after the current seclusion of contagion is lifted. I pray that my efforts will aid in the advancement of our Gentle Craft, putting the tools of technology in capable hands and keeping its role in due bounds. For all of us, I repeat the invocation I've used at the communication conferences I've held over the years:

> Grand Architect of the Universe, empower us to serve the Craft for the benefit of Mankind and each other, with the talents you have given us, and the will in our hearts, to learn and act in the noble spirit of peace and harmony.

So mote it be.

Bro. Ken JP Stuczynski
29 March 2020 A.D. / 4020 A.L.

HOW TO USE THIS BOOK

It is the author's hope that you, the reader, will find many useful, actionable items within the covers of "Webmastering the Craft". However, this is not a "how-to" book. By the time you've bought and read this, something in the digital world will have changed the game, perhaps significantly. That is why you should think about what you read here as methodology, a general wisdom that can be applied to whatever comes next. After all, human nature will not have changed, even if the speed and scope of our ability to communicate does transform – outwardly at least – the way we may live.

The first section covers various strategies and mechanics for communication and public relations. It is not primarily organized by particular channels (email, websites, social media, etc.), mainly because these are all part of a bigger picture. Without context and purposeful intent, focusing on the technology itself can cause the tail to wag the dog. Instead, there will be insights into specific technologies throughout.

An important note: Your circumstances may be unique, or at least different than someone else reading this. One Lodge or chapter may be rural; another may be in an urban metropolis. Demographics of age and the target audience of new members may be nothing alike. You will have to decide what is best to use and how to use it best.

Moreover, the principles here may be applied to *any*

organization, not just Masonic ones. Even if Masonry is espoused to being more than a social club, we are still social beings. These ideas and strategies are meant to address those temporal needs common to all organized groups.

The second section is about larger organizational structures, specifically those organized like Fraternal bodies with individual chapters, within districts, and all under a grand body. It outlays a model that may be used to coordinate communication within the larger culture, with particular emphasis on the district (or regional) level, as it is key to broadening the sharing of responsibility at all levels. Masonic terms will be used, but any reader may translate those into their own respective organizational language.

The third section is the complete collection of *Empire State Magazine* articles published by the author up to this time. These articles preach many of the lessons in this book, along with more esoteric musings related to how and when technology should be used in the Craft. They may be reprinted by permission of the author and the Grand Lodge of the State of New York, and any can be used as a presentation or for discussion at a Masonic gathering, if appropriate (no permission needed).

Lastly, you are encouraged to share what you have learned, not just from this book but from your own travels and practices. This tiny work is not at all exhaustive and will undoubtedly be expanded upon, perhaps with welcome contributions of your own.

DISCLAIMER

This book contains guidance for webmasters and others who serve the craft in digital media. Much of it is oriented specifically for the jurisdiction of the Grand Lodge of the State of New York. In general, these are meant to be explanatory rather than authoritative, and if in doubt how to use this information, consult your Grand Lodge through appropriate channels.

The author is publishing this work independently of his affiliation with The Grand Lodge of the State of New York or any other Masonic jurisdiction or body, and therefore no endorsement is implied.

THE QUARRY AND
THE WORK

CHAPTER 1

THE IMPORTANCE OF DIGITAL LITERACY

LITERACY AND MASONRY

Before we talk about the Digital Age, please indulge with me an esoteric musing: why is Masonic ritual so founded in the concept of mouth-to-ear? Only in recent times have most jurisdictions begun to use printed representations of ritual, and even then some jurisdictions only make enough copies to be used by District Deputies to ensure standard work. It is meant to be spoken from memory and not read. The most fundamental means of recognition are preserved as undescribed gestures and unwritten words rather than "lawful Masonic information", i.e., dues cards and passports.

In operative times, illiteracy was the norm. An apprentice would not be expected to do more than carry messages — they could not write or read them on their own. But as they proceeded to discover the trade secrets of their craft, some would have been taught a plethora of arts and sciences. They still needed symbols and passes to interact with workmen of varying literacy (and languages), but now they entered a new sphere of written knowledge. A master of stonework was likely to be one of the literate elite.

Let's go back farther.

Monasteries had preserved (hoarded) some fraction of ancient texts after the fall of Rome, but it was the intellectual pillaging of the Crusades that resurrected the foundational knowledge of workable arts — mathematics, astronomy, and advanced building sciences. The great cathedrals of Europe were not built by monks; the Age of Enlightenment that followed was filled with those "accepted" into Masonic Lodges, long before certain tavern Lodges converged in London in 1717. I do not think this is a coincidence.

Today's world is the reverse. Literacy is the norm. Handshakes and whispered words seem superfluous to dues cards. In my jurisdiction of New York, such credentials are now plastic with a scannable QR code that goes to a website or phone app, giving real-time status through databases talking to each other across the 'Net.

The thing to remember here – and throughout this book – is that we Masons do a lot of things that the World sees as archaic. But some of these things are not meant to be judged in that light. I would suggest our grips and signs retain their meaning in a profound rather than profane way: it is the communication between *spiritually* literate people in a spiritually illiterate world. And mouth-to-ear is even more important to human beings in a growingly impersonal digital world.

But we must give to Caesar what is Caesar's. We are inescapably *in* the world even if we aspire to not be *of* it. We cannot afford to be illiterate in the ways our current society demands, even if we are founded in timeless truths. If some Lodges and bodies are social and organizational dinosaurs with no concern for the needs and expectations of new generations, so be it. But Masonry has survived on the backs of those who have driven change in both the Craft and human civilization at large. And such men have always had to overcome the resistance of other members.

(On a personal note, I am no longer a spring chicken, and therefore likely to become less tolerant of change over the coming years. I hope I can overcome myself, or at least step aside, if I become such a hurdle.)

LITERACY TODAY

So what is literacy in its most basic form? It is the ability to communicate over time and space. In the old days, it meant you could write messages or books and have them read by other literate people, no matter how far away or into the future.

In my research as both a chaplain of an American Civil War descendant's group and (briefly) a reenactor, I learned that it was the company's chaplain who often wrote letters home for soldiers who could not read nor write. Everyone knew *someone* they could go to for such services.

And now we fast forward to a grandparent asking their grandchild to email a scanned document to a municipal office, or fill out a company's online form for some necessary purpose. They may forego seeing the faces of their own relatives out of embarrassment for not being able to use a cam, or simply go without that one thing they want to buy because they don't trust shopping online. Living like this is at best a litany of missed opportunities compared to what those around them have been doing for years. At worst these are harmless inconveniences – mostly to themselves – and are rightfully tolerated by those that love them.

The problem is when key people can't function in their required tasks. If you have a secretary who doesn't "do email", it slows down rocket-speed processes to that of a horse and buggy. (Long ago, my Grand Lodge made having email a requirement for being a District officer.) Even *one* officer without a computer or smartphone doubles the time it takes to communicate a message the rest can receive instantly.

But does this mean that we should ostracize or ignore those who are for modern intents and purposes "illiterate"? If a majority of members opt-out of modern life as the rest of us know it, that's alright. We can accommodate them. But *somebody* needs to be able to handle digital-based activities.

Remember the hypothetical secretary above? Delegating digital communications reduces drag on the process considerably. Having even a single officer or point man to handle digital newsletters, calendars, and other such processes makes all the difference. And if this person is not an elected or appointed officer, they must work closely with those who are. This will be explained much more in a later section, but the upshot is that if digital communications are dependent on a small number of people, they cannot be treated as a separate committee, but as integral to most or all activities.

LESSONS LEARNED

What are the lessons here? First, there's no harm in meeting on the level with those clinging to the past, so long as we don't neglect accommodating the future. Secondly, everyone doesn't have to be literate in the modern sense, so long as someone is, and they are tasked to do those things. Also, we should consider literacy as an important qualification for some leadership positions. Lastly, if digital actions are delegated to one or a few people, those individuals must be integrated into the team and its leadership.

CHAPTER 2

MASONIC JURISPRUDENCE

There are plenty of technologies to take advantage of by any organization. However, Masonry is somewhat unique in that we draw extra lines between what is public and what is private. This is not just an affectation of traditions with lost origins — there are real reasons this is so. The lines may be drawn a bit differently by different jurisdictions, and there are some grey areas due to special cases and new dilemmas caused by the emergence of new technologies. But old principles hold true, waiting to be interpreted and applied to all things new.

Our obligation is concerning ritual and modes of recognition, and we have a confidence between brothers regarding the discussions in closed Lodge — and general internal matters, of course. But as far as details, a Brother (from any jurisdiction regular to us) should know SOME things, such as the time, place, attire, cost, and anything else necessary to attend Lodge. Without good reason, these should not be kept a "secret".

The most practical and easy rule to follow is that anything public in the real world can be public in digital media (websites, social media, etc.), giving considerations of positive public relations throughout.

We should assume anything communicated without the restriction of membership to a media channel (closed groups on LinkedIn and Facebook for example) is public and govern ourselves accordingly. If we announce a meeting and someone shows up who shouldn't, that's what we have a Tyler for!

Let's clarify a bit more. We make public references all the time to tools and expressions used in ritual, on posters, in open meetings and mixed crowds, etc., but in such a way the secrets are hidden in plain sight. Non-Masons just won't get the reference or nuance, or even notice it at all. You could use an image of a winding staircase of a certain number of steps on a Facebook post or even a highway billboard, but an explanation would never be rendered except between Brethren we know so to be.

A final consideration would be discussing Lodge affairs even in a closed group. There is something in our obligations not to create separate discussions and groups separate from the Lodge leadership. I would recommend that any such closed group online be moderated officially by the WM or his representative to avoid even the appearance of impropriety.

In any reasonable and logical discussion of "appropriate" website content, there will be the good, the bad, and the ugly (white, black, and grey areas, respectively). This is especially true with the addition of rapid changes in technology.

At this point, I will remind the reader that I am not an authority on Masonic jurisprudence and recommend you apply personal and Masonic "common sense".

In theory, and hopefully, as part of good taste and security concerns, there are things that should readily raise red flags when considering content on the Web. Items like personal and financial information come to mind. Such things may wind up inside meeting minutes as a normal course of business, but that is where they would stay.

STRICTLY PROHIBITED

Like the old Vegas adage, "What happens in Lodge, Stays in Lodge!"

In 2015, the Grand Master of my jurisdiction gave an edict regarding the recording of Lodge proceedings, applying longstanding principles and expectations anew. One of the most common questions it answered was if the minutes of a Communication could be mailed or emailed to members.

The answer is no. Masonic Constitutional law is very clear on that, along with a 1936 decision of Grand Lodge (codified in §798). Lodge Meeting Minutes may not be distributed in any way outside of the Lodge. Members may view them at the secretary's desk, *period!* This includes publishing such transactions on a Lodge website, even if password protected. Just don't.

The spirit of the law is this: the information must not be available in such a way it could be visible to others in any way. It is a reference for the Secretary and Master, and may be shown to a member, but must not be in a form they can take with them. The moment it is photographed or can be seen on a screen in any form, it is no longer assured to be private.

This isn't paranoia. If you are a Mason you should understand why we do this — to allow us to do our Work without the judgment of those who are not privy to our ways and understandings. But there is a further reason. Someone's membership in the Craft is their own business, and they have a right to not have it known they are, or have even applied to become, a Mason.

What some of us do not know or appreciate is that in some places and circumstances there are still negative consequences for being a Mason. A law enforcement detective, a teacher, a politician or a famous person — these may all want to not be known in their whereabouts. In the worst case, there are Brothers in my jurisdiction whose families back home would be in danger if their respective governments found out they were Masons.

PHOTOGRAPHS

This is also why we need to respect all requests to not be in photographs. It is unreasonable in my opinion to get explicit consent from each person in every group photo. We live in a world where even our food is famous on social media before we eat it. If there is an issue, I would hope the member would make their preference known, as it is the exception and no longer the rule. But alas, I am not a lawyer …

Photographs are also inappropriate, for the most part, during a tyled meeting. It is a private space, a sanctuary. The exception is if someone is receiving an award, but photos can always be deferred until closing. One admonition that may or may not be universal is that the Great Lights should not be photographed, or in any photograph, when open.

GREY AREAS

This is the area where your Masonic education and common sense come into play.

If Lodge meeting minutes are stored (by the Secretary) in the Cloud, I would recommend strict diligence in not allowing anyone access except through the secretary as prudent or necessary between officers. In other words, if a cloud-based storage medium is used as a locked-down storage medium, it's no different than a drawer or the secretary's briefcase or laptop. Just be aware that if made available in any but these most limited respects, it is arguably the same as publishing or mailing.

A trestle board "activity summary" posting or mailing that can easily be construed as, or are proven to be "the" minutes of a meeting but only reorganized or reworded leans heavily into the black side of grey. This is not to say that an aptly constructed, guarded summary of the work activities, programs conducted, or guest speakers in attendance, should not be included in a public forum such as a website.

But each Grand Lodge may rule, or have already ruled, on these matters. Be as aware as you can of any laws, bylaws, edicts, and rules related to such things. Obey the orders of the Worshipful Master. Challenge and seek the reasons we do things the way we do, but consider complying with tradition if it keeps the peace. As times change, I am confident we will find our rightful paths.

A WORD ABOUT MASONIC "SECRETS"

What is or is not a secret, and therefore should or should not appear publicly, digitally or otherwise, is outside the scope of this book. There is a broad continuum of opinion on this subject, where some say it refers only to modes of recognition, and others anything uttered in ritual.

I will only say this, for what it is worth: Many public books speak of our lectures and rituals in some detail, even quoting things that might be cyphered in our practice books. But as our symbols are not hidden from the outside world, a phrase or passage is not revealed just because it is readable — the uninitiated will not know what it is referring to. It is merely poetry or fancy language.

But in the end, we must all contemplate our obligations and examine our consciences, so I leave it ultimately to the reader.

CHAPTER 3

PUBLIC RELATIONS

The Internet is by far the most used form of media and information nearly everywhere in the world. To ignore or be absent from it is almost not to exist. That doesn't make it the best medium for everything, or one everyone in your audience uses, but there is now a whole generation that does not remember life before email and the Web. And we're even past that. Social media is ground zero for everyday contact with friends and the world.

Your Lodge may or may not have a website, but posting updates and news that way is not as important as it once was. An announcement there by itself means little. People don't know there is something new on a site unless you have am email or text notification system in place people can subscribe to. More often, the site is for hosting the content you want people to know and it is promoted by a link in other places or email. Some sites are set to automatically post new content to social media.

Every Lodge must decide if they want or need a website (instead of information about their Lodge on the district site or some other place). But if you have one, it is visible to the world, so it must look professional and be reasonably up to date. Intentionally or not, your web presence is part of your face to the world, so again, don't think only of the older brethren who don't care how it looks or if it works.

Email lists are a great way to get the word out, particularly within the Craft. They can be public where anyone can subscribe, or just members. Whichever it is, govern yourself accordingly. Consider having a list that goes out to news outlets (similar to fax lists in days of yore), but keep in mind that there is no substitute for personal emails when building relationships with local press.

News outlets prefer press releases via email, for the simple reason they can copy and paste the information without retyping it. However, cut-and-paste shifts the responsibility of proofreading more to the writer than the reporter, and sometimes whole releases are posted as is, mistakes and all.

Posting to social media is often the best way to get a broad response from those in (and interested in) Masonic events and activities. Even photos after the fact can be tagged and shared among Masons, their friends and family, and the people those people know. Social media is so important today that it is better to go without a website than not have a Facebook page people can like and share.

Whatever outlets or methods you choose, stick with common formats. Don't use a picture of text when you can use text unless you do both, the accompany the text with the printable poster or brochure. Avoid sending things in Word or other proprietary formats — plain text is best. Others may not be able to open such files and if they can, they will probably have to format it their way if republishing anyway. For photos, it is better to be too large than too small. As long as it isn't too big to send (many email systems have a 5 Mb limit), it can be resized on the other end. (You can't make images larger without losing quality.)

Also, know that Grand Lodge exists to serve its Lodges, and their channels of distribution are available to you. In New York, for example, if you want something published in the *Empire State Mason* magazine, on NYMasons.Org, or in the *Hiram's Highlights* email newsletter, send it to anyone on the Communications Committee. (You can also join and post information on the Atholl discussion group previously mentioned, though Yahoo is slowly shutting its groups down.) These platforms reach most or all Masons in that jurisdiction, as well as Masons and others who follow them around the world.

Lastly, remember that when dealing with digital communication and media, the basic rules have not changed. The five W's, proper grammar, and common sense still prevail. What you do online reflects on your Lodge and the Craft as a whole. If in doubt, consult your leadership. Your Grand Lodge may have a PR manual or people appointed to assist in such things. Ask and ye shall receive.

CHAPTER 4

A MATTER OF STYLE

The term "style" here refers to the various aspect of how something is written. It is all about standards and consistency — not just grammar or tone or verb tense, but a standardization of abbreviations, capitalizations, and in some cases, even fonts, italicization, etc..

There are more and less common conventions. For example, the word Lodge (capitalized) refers to a speculative group of Brothers, their meeting, or meeting place, whereas lodge (uncapitalized) is proper for an operative guild-body of stonemasons (or a building so-called unrelated to Freemasonry). Brother as a title should be capitalized and is usually also done so in the formal plural, Brethren. Officer titles are usually capitalized, as are central Masonic emblems and terms, such as Apron, Square, Degree, and Charter.

Tiler is preferred to Tyler, at least in American jurisdictions it seems, but I prefer the latter, and my also being the publisher of this book, I can get my way here.

Sometimes there are contradictions. A common style guide uses '#' for the number of a Lodge, whereas the formal rule according to my jurisdiction is 'No.', i.e. "West Seneca Lodge No.1111" versus "West Seneca Lodge #1111." The uncommon preference of having no space before the number is my own.

TITLES

The most common inconsistency is that of titles. It is common and acceptable to simplify titles by omitting the difficult-to-format "three dots" character. In other words, "M∴W∴" and "R∴W∴" are rendered "MW" and "RW" respectively, except perhaps in historical quotations where they may be preserved. This is standard for publications such as the "Empire State Magazine", in which I have been a regular columnist. What I find awkward is "tricks" to make the three dots, such as periods and apostrophes or a colon and period.

To my preference, I abbreviate the title Worshipful as "Wor."; the title Brother is generally spelled out at the beginning of sentences and abbreviated otherwise as "Bro.". In hasty and informal communications, I prefer a single slash, (i.e. "RW/Brother"), rather than the common two (i.e. R/W/ Williams). To each their own, so long as they are consistent.

TECHNICAL NOTE ON THE THREE DOTS CHARACTER

The three dots character ('∴') can be found as a special character and cut and pasted as necessary for use in printed material. However, the character has been shown to cause issues with online RSS feeds, and should probably be avoided, especially in blog posts.

THE QUARRY PROJECT STYLE GUIDE

As you may have guessed by now, there is not a style guidpe. However, the following is my recommendation to the Craft — The Quarry Project Style Guide. It is an effort to establish a consistent style for Masonic writers and publishers in the

United States for periodicals, books, and websites. The guide gives direction on source citation, spelling, punctuation, capitalization, numbers, abbreviations, quotations, tables, figures, and format.

This guide is based on "Chicago style," as presented in Kate L. Turabian's *A Manual for Writers of Research Papers, Theses, and Dissertations, 8th ed.* Users of the Quarry Project Style Guide will need to reference that work. (Be sure not to confuse it with Turabian's lower-level book, *Student's Guide to Writing College Papers*.) Information can be found at TheQuarryProject.Com.

INTERNAL VS. EXTERNAL STYLE

A public relations guide, if your jurisdiction has one, can inform you of the proper language used in public information releases and advertising, but here are a few hints, starting with a longstanding rule about the mail: Masonic titles are NEVER used when addressing an envelope for mailing. Whatever is in the letter is fine, and there's nothing wrong with a return address of a Masonic Organization, but it is a matter of privacy that the receiver is not assumed to be known as a Mason or their position.

Some of this is common sense if you think about it. Though officer titles (but not rank) are acceptable according to one of my jurisdictions manuals, the title "Worshipful Master" may be confusing in a news article, rather than say, "presiding officer". Masonic *Hall* may be preferable to Masonic *Temple* for the same reason. If not already prohibited, do not use the terms Initiated, Passed, or Raised.

Internally, there is much more freedom to use words as

we use them, of course. Good advice would probably be to remember the new name you were given.

CHAPTER 5

THE RIGHT CHANNELS THE RIGHT WAY

There are more ways to communicate than ever. Our Masonic "Communications" up to the present time have always been in person. But how we keep brothers informed with regards to Lodge matters can take many forms, physical and digital. How we make ourselves known in the community, as well as publicizing public events, can be equally diverse.

This chapter is mostly regarding internal communications. Some Lodges are so bad at communicating, if you don't show up for Lodge, you don't know what's going on. I also hear murmurings that it's the Brother's fault — if they wanted to know, they should have been there. But we know it is simply not possible for all members to be at every communication, even if they wanted to.

Should Brothers just spread the word? At one time, that might have happened in natural course, with the Brethren all living in the same town, going to the same church, and working for the same few local companies. Those days are gone. Community Lodges in that old sense rarely or ever exist anymore. Most people live, work, and go to church in completely different places. And the 9-to-5, 40-hour workweek is no longer the norm. It is unrealistic for Brothers not at Lodge to know what is going on, and they shouldn't have to call their friends or the Master every time they miss a meeting.

And what of out-of-town or infirmed members? I can personally attest that some, apart from receiving yearly dues notices, have not heard from their Lodge or chapter or any other Masonic organization for years. In my opinion, that is insane and an affront to our obligations.

With that in mind, let's review the many ways we *can* keep in touch.

PHONE AND PHONE TREES

The most direct method of communication other than in-person is the humble phone call. This is most appropriate for personal business, health checks and condolences, and important notices when you want to be certain someone is personally notified.

Phone trees spread out the responsibility and allow for such personal contact. The current Master of my Mother Lodge asks the officers to each call a portion of the roster for such things as funerary services and to invite them to special events. This sort of thing cannot be overdone, except by exceeding the cable tows of those making the calls. And in the process, we discover what we don't know about Brothers we haven't seen.

As Master of a Lodge, and now Sovereign Prince of my local Valley, I made it a point to reach out to every Brother in our membership early in my term. I considered it as taking inventory of our "living stones" without which we have no purpose or means to do our Work. In that course, I discovered deceased brothers as well as those in need of visitation or relief; many could not be reached because their contact information was outdated.

NO BROTHER LEFT BEHIND

Keeping contact records updated is everyone's responsibility. The Brother should – if they are able – inform the secretary of changes in phone and address, as well as email (if they check it). But nothing can be assumed. People move, sometimes not of their own volition (committed to nursing homes, for example), and sometimes the family does not make the Lodge aware of a Brother's passing.

The solution to keeping in touch lies with phone calls more than any other method. You know if you talked with someone, whereas it may be unclear if they received a letter or email or robocall.

If you get a disconnected number or find yourself always leaving a message on a generic answering machine, it could be the wrong number. There are two things you can do: attempt to look up their information (I use WhitePages.Com); send them a postcard.

When I leave a message on a voice mail or answering machine that doesn't tell me I have reached the intended person, I add something like, "If this is *not* the right number, please call and let me know." Many people have done the courtesy of doing so, and then I know to begin the search.

A postcard should ask them to contact you, explaining you do not have a working number for them. Make sure it's personable. Say who you are and how to reach you. It may come back undeliverable and you know it's the wrong address. Many times I have gotten calls from family, and even an attorney handling a brother's estate. If it does not come back and no one calls, it still may not have gotten to the person intended.

If all else fails to find a Brother (or widow), you can put the call out to the Brethren, as someone may know. If within reach, you could visit their last known address in person. Online searches may save you a lot of that trouble. It's just a matter of effort Brothers are willing to make.

ROBOCALLS

Automated calls (such as **PhoneVite.Com**) can be reasonable in cost and get the job done. You can set up when you want them sent, and keep track of who received them and how long they listened to the message. This method is good for urgent notices where other methods might not reach members in time, but also for general reminders that do not have a lot of detail or require a personal touch.

Because it is not "captured", repeating the message will give the listener time to make a note of details they may wish to remember. You can set the call to be from your number so it will more likely be answered. (My wife lets all calls from the Sister who sends these go to voicemail so she can save and listen to them when she can make a note.)

An important note: Be aware of brothers who opt out of receiving your robocalls. If they continue to receive them they can have a legal claim against you for quite a sum of damages. The main reason you may knowingly not comply is if you refresh the list from Lodge records every year or so, as such members may inadvertently be readded. Make a special note of those.

Otherwise, robocalls are easy. But keep in mind that they are not useful as a reference by a Brother for detailed information. That is what print is for, be it made of pulp or pixels.

"SNAIL" MAIL

Postal mail went from the main means of communication to an anachronism to something special. Because of its infrequent use (other than junk mail), it can really stand out. A simple physical invitation to an event – even a postcard – can be perceived as acknowledging the receiver's importance to the sender.

More importantly, a physical note can be kept in sight, such as on one's desk, the family fridge, or in the ... ahem ... auxiliary reading room. Afterward, it can be kept in a scrapbook, or as a keepsake in general.

EMAIL

Emails are instantly receivable letters. They may sometimes be terse, but good form is still relevant. In a cold contact, address the person with a greeting and close with your name and information. Don't YELL in all caps, or as my wife says, "use your indoor font size". Avoid colors or special formatting.

Make sure your email program has your real name in the basic 'from' field, especially if your email address is something like 'chowhound99' and unidentifiable to recipients who don't know you. An email signature is ideal, but not visible at a glance or when being CC'd.

So what is the strength of email? Email is ideal to send detailed information that the viewer can then refer to later, or copy and paste as needed. You can also send contact information (vCards) and calendar event files that the receiver can easily integrate into their own address book or calendar.

It's especially good if you need to document a conversation for accuracy and accountability. It can get a bit tedious if many people are CC'ing each other back and forth, but it's better than not communicating between real-time meetings.

(E-)NEWSLETTERS, BULLETINS, AND BULK EMAILS

A newsletter or bulletin should be a thing of beauty. Alright, maybe not high art, but something better than a single-column printout of paragraphs. A graphic or two can go a long way, but avoid unnecessary visual cliches. Use columns (less important in e-newsletters) to make it easier to read, and boxes to set aside content you want to draw particular attention to.

Physical newsletters, like other postal correspondence, can be closed by a sticker to save buying envelopes. However, some Masonic information requires a *sealed* envelope, such as the names of candidates. Your Secretary should know the jurisprudence on such matters.

Digital versions of a print newsletter should be a PDF file, as it is the only universally accessible format. (Not everyone has Microsoft Word.) There are free converters online if you don't already have that capability.

E-Newsletters, other than emailing a PDF of a print version, can be done on any number of free or inexpensive email campaign platforms. Two common ones are **Mailchimp** and **Constant Contact**. The learning curve may be a bit much, but it is worth it.

The benefits of an e-newsletter are many: you can automate subscription; you know who opened the email; you can archive sent messages and save campaigns as templates.

Very importantly, they have built-in compliance with federal laws. If you aren't familiar with the CAN-SPAM Act of 2003, I recommend looking it up. Here are the highlights: any bulk email sent must make a subject header that is not misleading; it must give a link or other way to unsubscribe (which must be honored); it must give a mailing address of the sender.

You can manually add the necessary information using a standard email program if you go that route. But there is one more pitfall to avoid. Do NOT list the recipients in the CC field, unless you are sure everyone on the list consents to everyone else seeing their email address. Use BCC instead.

PRIVATE EMAIL LISTS (LISTSERVS)

Group email lists are less common than when listservs were all the rage (I'm telling my age now) but can be very useful. A single email address can be set up so that any email sent to it goes out to a list of addresses, such as the officers of a Lodge. This list can be changed with each passing election or other change, and makes it easy to reach pertinent people without digging through an address book hoping you didn't miss someone.

Of course, this is predicated on all or nearly all those to be reached having an email address, one that they check often. Depending on how replies are handled and if everyone can post, it can get messy like a group email chain or group text. And it can be set to only allow certain people's messages through and not everyone. It often has an archive feature.

Then there is the unofficial yet officially promoted and moderated "Atholl 1781" Yahoo group. Also known as the "Grand Lodge of New York Information Net," it is a web-based example of this. Sort of. Discussion or "newsgroups" (again showing my age) are forums for posting messages, almost like a closed social media group, but because people subscribe to receive messages (or digests of messages), it acts like an email mailing list. Many get use out of it, but I do not recommend using this fading medium.

TEXTS AND PMS

My daughter rarely answers the phone if I call her; she will answer a text right away if humanly possible. There are many reasons for this being typical, from generational preferences and workplace policies to social mores. Email was the dominant form of communication since Y2K but has been displaced by texts for a few years now. Clients often contact me via Facebook messenger, and I hear that it is not uncommon.

And some people prefer to receive all sorts of updates through text. From pizza and pharmacy pickups to coupons and appointment reminders, more and more people count on them for everyday things. Is the Craft using them? Absolutely.

Group texts can be annoying, but there are systems available for people to subscribe to text messages. Some websites can be set up for this, or it can be handled privately. The value is that something can be broadcast on that medium, not as a conversation, but as a notice. This is perfect for last-minute changes in events, such as a weather-related cancellation or some emergency circumstance.

PHONE APPS

Phone apps are simply little programs on a smartphone or other mobile device. They can provide up-to-date information by communicating with some other system, such as that of a weather service or a database of members in good standing. I don't develop apps myself, but understand that they can be very simple or as complex as you need, with respective costs and time to develop.

At least one Masonic district in my jurisdiction has its own phone app, tying in Lodge information, events, and contacts. The "Our Lodge Page" app, developed by the same company that provides our jurisdiction with membership record management, allows users access to contact information for Brothers of any Lodge hey are a member of, as well as events entered into the system by Lodges that choose to participate.

The **Amity** app is something to check out. They are working with many Grand Lodges to provide Masons with information on Lodges around the world, and can be connected to a membership system to confirm the standing of a member when presenting a permanent dues card with a QR code. This is used in many Masonic jurisdictions and Masonic bodies already.

TOUCHING ALL BASES

The most important thing to remember is that there is no one best method to communicate. The urgency and nature of the message should determine what method is appropriate, but more so, we must meet every Brother on their own level. Different generations prefer different media, and some don't have access to all forms of it.

A balance must be struck. For example, when I was Master, the plan I followed was a physical, mailed bulletin every two months with interstitial email newsletters, along with the occasional robocall reminder. Vital information was mailed, and "bonus" material was sent in between. Email was used for documented instruction and general messages. Phone calls were for personal discussions and reaching out to long lost Brothers. Even though some Brothers were more aware because of their own choice of digital literacy, no Brother was left in the dark.

CHAPTER 6

DON'T WORRY, BE SOCIAL

N o "channel" of the Web is more contentious than social media. By default, we think of Facebook, but it applies to any website that allows people to create an account under their name, make connections with other users, and share (and re-chare) content. For some it's just part of life or an addiction; to others, it's a chore, or seen as a cesspool of bad behavior. It is a reflection of us, minimally filtered — the good, the bad, and the ugly of everyday people's lives. And giving everyone a voice and contact with each other anywhere, all the time, puts it on a scale of changing human existence akin to the invention of writing itself.

You may not like it, but ignore it at your own peril. If a group doesn't have a "page" or "group" one on some social media site, someone will eventually create one without permission. Then you have to wrest control of it or hope it can be shut down once it becomes problematic, or conflicts with an account you *do* control (the "official" one).

To control, or at least guide that part of your online presence, you have to participate in at least some small way. You can create an account to claim your "brand" and not use it. Just be sure that if you do commit to occasional updates, don't let it go untended too long, or it will look abandoned, and, like an outdated website, will make people think you are "out of business".

FACEBOOK

Social media platforms are communities, and not small ones. If Facebook was a country, by active users alone it would be by far the largest country in the world by population. If you think your members don't use it because they are too old, think again. And if a member doesn't use it, their spouse probably does. Of course, your demographics may dictate how important it is. Retired women are a fast-growing segment; young adults now favor other sites but still have a presence there. The sweet spot for social media is Millenials; for Facebook, it is more Generation X.

The important thing to know about Facebook is that it has "Pages" and "Groups" and lots of controls for privacy and moderation. These are not personal "Profiles" — don't set them up like you would a person. It's against their policy and could be terminated at any time. If you can become "friends" with a company or group, they did it wrong.

PAGES AND GROUPS

A **Facebook Page** is a different kind of profile, belonging to a business, organization, or cause. It is for disseminating related information. Users can "like" a Page and "follow" it so when something is posted, they may (but not always) see it in their news feed, or "wall", just like posts by friends. It is usually set up by someone with a user account, and that person becomes the "admin" — the administrator with rights to control the page. Permissions can be set to allow or not allow other people to post, or assign other people to manage the page. An important note regarding privacy: the profile of the person who sets it up or manages it is *not* visibly tied to the Page. Unless they choose to, anything they do on that page will be done under the name of the page, not their own.

A **Facebook Group** is like a Page, but oriented toward discussion. People can "join" a group, and activity therein may appear on their news feed. Every post can be a topic that others can freely comment on and reply to others' comments. It may be set up by a user through their profile, and anyone who is a member can be made by that person to be an admin or moderator. It can be closed or open, meaning you can control who is a member and can see the content inside.

Either way, a Page or Group should be clear as to what organization it represents and link back to a website if you have one. Know who administrates it and make sure decorum is enforced. Don't make it a free-for-all, but don't restrict it so much people cannot participate or share content. You'd be surprised how rare it is that people post inappropriate things, and even then it's more often blatant self-promotion or misunderstood comment than something egregious. Encourage Masonic gentility and tolerance.

My recommendation for most membership-oriented organizations is to have a *public* Facebook Page and a *private*, members-only Facebook Group. Many things will be posted on both, but the latter allows for more freedom of discussion, and notices of strictly internal affairs.

TWITTER

Famous for its use by the rich, famous, and powerful, it could be described as microblogging or a simplified version of Facebook. An account can represent a person, brand, company, or organization. Each post, or "tweet." currently has a 280-character limit, so it is more of a shout-out to those following an account, the same way a short text compares to a long email. Some people live for Twitter, and it's followed closely by journalists. It's a useful but not essential channel, but if done, be sure to understand how it works and do it right. Following connected organizations and getting followed back is the name of the game.

LINKEDIN

LinkedIn is only mentioned here because of its common use. It has company profiles, but is predominantly about individual professionals networking — a sort of Facebook for business people to promote their careers. I do not find it useful at this time for organizations to include it in their social media presence, but the leaders of the Craft may want to have their own profiles to interface with other leaders as a personal choice.

On the other hand, a Square Club or Lodge with profession-based membership qualifications may want to have its own LinkedIn Group as an alternative to (or in addition to) Facebook.

VIDEO CHANNELS

A video channel is a great way to be in the game these days.

Not only does it center on videos such as events and personal messages from leadership, it acts as a repository for them. **YouTube** is the dominant player, and when the description has links to your website and other channels, it gives you search engine hugs from its owner, Google. You can also post videos directly on Facebook, or share them on other social media channels. Videos can also be embedded in your website rather than be stored there.

INSTAGRAM, PINTEREST, SNAPCHAT, TIKTOK, ETC.

There are seemingly endless Socal Media platforms, each with their unique capabilities, strengths, limitations, and user demographics. For sharing photos, videos, and photo albums, **Instagram** is hot at the moment, and many businesses have a presence there. **Pinterest** is mostly for sharing and categorizing images, and TikTok is the latest craze in sharing mini-videos, but at this time mostly for teens and young adults.

Some of these are less professional and more personal than others; some have larger audiences than others if that is important to you. Whatever ones you choose, do them with decorum befitting our values.

Again, social media cannot be ignored. It is so important that for a small organization or certain types of business, it would be better to have a Facebook Page than a website if you have to choose between them. And this is coming from someone who sells websites for a living.

There are so many nuances to managing social media platforms and their diverse and ever-changing processes and quirks. That's unavoidable and requires patience. Have a plan, decide a level of dedication to it you can be consistent with, and make sure it stays done. If you can't stomach it, find someone who can and delegate.

CHAPTER 7

MEETINGS, COLLABORATION, AND THE CLOUD

MEETINGS & EVENTS

The term "meeting" can mean many things for a Masonic organization. It can mean a formal communication (regular or specially summonsed), an officers meeting, or a public ("open") gathering, presentation, or other event.

The prohibitions against broadcasting or recording a tyled communication should not have to be explained. As things stand, to participate in a tyled meeting, you must be physically present.

For all other events, you may consider the option of a telepresence or videoconference. Video calls are not uncommon, and software like **Skype** has allowed people to be virtually present at weddings, or the birth of their child while deployed half a world away. Nursing homes can use these technologies to be a little more "in-person" than a phone call for distant relatives. Things like **Facetime** are catching on, finally, after generations of longing for Dick Tracey's watch.

If it's only a person or two, someone on speakerphone or a video app will suffice. For more, teleconferencing does the job. Although around for years, teleconferencing no longer requires special equipment — it is hosted over the Web and accessible to anyone with a connection. Videoconferencing is just as easy, and most systems allow for call-ins even if they cannot or do not wish to do video.

The ones I see out there the most are **GoToMeeting** and **Zoom**. (The latter is being promoted widely for use within our jurisdiction during this trying time). They both have free plans, with limits of time, number of participants, and features, but most have desktop sharing and chat features. What you want to consider in choosing a meeting platform or paid packages are other the features you may wish to use, such as recording the meeting or holding a full-blown seminar with thousands of participants and viewers. But for most Masonic events, even the free versions stand up pretty tall.

To use these requires a basic comfort with technology. Anyone can call in. Those wishing to utilize video may be prompted to install a program on their computer or device, but that is usually very easy and seamless to the process of using it.

The real question to be asked is why we aren't already widely using these technologies? It's almost impossible to schedule off-night meetings with more than a few people due to schedule conflicts. If people could call in, traffic is no longer an issue. Things like being on the road or babysitting is less of an issue. Participation would skyrocket for some leadership teams.

More importantly, we could involve out-of-town members in committees, invite them to join us virtually for dinner, or a labor-to-refreshment presentation. We could do this on a regular basis, and it would mean a LOT to some brothers who are infirmed or a thousand miles away.

TYLED AT A DISTANCE?

I can think of one way it may be appropriate to use a virtual presence for a tyled Communication, if it hasn't already been done. There are times two or more Lodges can jointly conduct a Communication, and there is a process for that. (Both Masters declare the Lodge open and closed, etc..) And like any other Communication while at work, there is a symbolic barrier between the Lodge room and the outside world. But what if *two tyled spaces* are magically connected? If the other end of an encrypted video stream is also kept apart from the world, I would suggest it is not in breach of our privacy, secrecy, or sacred space. This application of the spirit of our ritual could open up a new world of universal fraternal relations.

COLLABORATION

Tools for collaboration have been around forever. A posted trestleboard for the building of an edifice is a perfect and apt example. A shared bulletin board is a somewhat modern version of that. But now people use private, team-based, online forums to share notes, documents, and checklists. Whole workflows are set and followed.

Web-based platforms like **Basecamp** and **Trello** can have very different approaches and interfaces to achieve this, but the concept is the same. Like other software mentioned here, some are free with paid versions offering advanced features and fewer limitations.

But there's a simpler, less fancy way to do much of this — file sharing on the cloud.

THE CLOUD

The "Cloud" refers to little more than file systems online. Just like folders on your computer, you can store files, such as notes, manuals, graphics, etc.. The difference is that being online you can share these files with other people, and on some systems, multiple users can make comments and edits in real-time. It is also a way to back up files, such as photos from your phone, or work documents you wish to access when out of the office.

Cloud platforms allow a user to set files and folders to private, or give limited permissions (to view or upload only) for specific users or the public. There are many free cloud services, usually with paid versions, the difference being space. Some of the more popular cloud platforms are **Dropbox**, **Google Drive**, and **Microsoft's OneDrive**. The last two are connected with user accounts that have other services, such as calendar, email, and programs to edit documents.

There are many uses for the Cloud. For example, you can put a "Newsletters" link on a site that goes to a folder where another member can upload them as they are produced, creating a downloadable repository. Internally, you can keep team plans, instructions, and other documents for reference or editing.

And even though it's best to commit to and carry over one system from year to year, there is flexibility in even that — most cloud and collaboration software can integrate with each other.

CHAPTER 8

ALL ABOUT WEBSITES

A whole chapter is devoted to this for two reasons: it is squarely within the wheelhouse of my usual vocation; it is a foundation of an organization's web presence.

However, a website doesn't fulfill the same role as it used to. It used to be the be-all, end-all of a web presence. Since the dawn of social media, your presence (and "brand") is part of a larger web-wide discussion that will happen with or without your website (and control).

Think of a website as the foundation of a house. You don't live in the basement. Social media is the living room, where all the parties are held. But the foundation gives it a sturdy structure you can use to keep things in place — a "home base" for official information that you control, where you can send people to from these other channels, and from this central hub out to all your channels, be it Facebook, YouTube, Instagram, whatever.

What does this mean for your website? It means you are not under any pressure to constantly update it with new information, though you can. Most of your updates – your photos from the last event, your fundraiser fliers, your musings and messages – belong on social media where it has a natural reach and can be most easily shared. More on that later.

THE BASICS

In a later section, we will discuss in detail about choosing a webmaster, or even one person who will set up the site and another who will manage it. As modern sites are more CMSs (Content Management Systems) with WYSIWYG (What-You-See-Is-What-You-Get) interfaces, anyone with word processing skills and a bit of sense can maintain a website. And with some services and tools, no back-end coder is needed at all.

At this point, I must do some intermediate education about the web. The "World Wide Web" is the part of the Internet (the network of computers connected around the world) that supports websites. Websites, for the most part, are just files of pictures and words that connect to other pictures and words (links), forming a "web" on information of all kinds.

Websites, being mostly a bunch of files, have to have a place where those files sit and can be accessed from all the other computers on the 'Net, and to devices like yours who tap into it. In other words, you have to rent space somewhere that can accommodate those files and accept traffic (answer requests by the program people use to surf the web). For example, the files to my websites (including NYMasons.Org) are currently on a computer in a data center in Michigan. I pay them monthly to keep the lights on and the pipes open.

A Domain name is the "dot com", "dot org", or "dot whatever" that points to that datacenter, computer, and ultimately the files that make up your website. If you change where your website is hosted (where the files are), you simply change where the name is pointed. Note that at one time '.ORG' was reserved for 501(c) not-for-profits; today, anyone can use them, but sometimes '.COM' is easier for people to remember — the choice is a preference rather than a rule.

So you have three things going on here, all of which costs money (or are donated): you must pay a yearly fee for the name (i.e. "YourLodge123.Org"); you must pay someone to host the site, either monthly or yearly; someone must build it. Some companies may provide all three. The important thing is to make sure everything is in your organization's name and you have access to control these assets.

THE DESIGN PROCESS

As I make my living from setting up websites, I am a bit prejudiced against DIY (Do-It-Yourself) website platforms, such as **Wix** or **Weebly**. These enable anyone to potentially set up a decent-looking site. However, choices are limited, and the site isn't portable if you want someone else to host it.

I recommend using **WordPress**, as it is practically a standard for all websites today, and there is a vast community for support and options. To be clear, I don't mean having a site on WordPress.Com, though there's nothing wrong with making that work — I mean using that software on your own hosting server, where you have almost endless control.

However you do it, don't design by committee. This doesn't mean everyone shouldn't provide input. It means they should not be allowed to decide as a group how to execute all the various details as a whole. If it is not tightly brought together by ONE cook in the kitchen, the site will appear schizophrenic, or a circus of conflicting ideas. The fewer cooks in the kitchen the better, and they will have to answer to the head chef anyway.

If *you* are the guy in charge of having the site built (by yourself or another), here is some extra advice: people know *if* they like or don't like something, but will give you vague ideas of what they mean. The truth is they don't know most of the time, and you have to play forensics by showing them examples of other sites to drill down into what they do and do not want or like specifically. That may require some patience.

STANDARDS AND BEST PRACTICES

Before I became a member of the Communications Committee of my Grand Lodge, there were a few attempts at standards and rules for websites. They were never widely known, followed, or enforced, and didn't age well. But as progress was made in my tenure, I made it clear I was only opening the lid of a jar that my predecessors had worked hard to loosen for me.

My approach was that Grand Lodge should not micromanage such an amorphous task from above, but provide only the most necessary restrictions (based on existing Masonic jurisprudence), and encourage and educate for good practices. Each Lodge has its own unique identity and needs, and therefore no one can know better than them how to proceed.

The most important thing is to make a site visitor-centric. Yes, you want to push out particular information and initiatives, but think more in terms of providing information people will need. And that means up to THREE target audiences — the general public (PR and membership development), members of the Fraternity (news and education), and leadership (administrative resources). If you can pull off satisfying each of those target audiences without making them jump through hoops, you succeeded.

The term standards here means bare minimum suggested requirements. They are based almost wholly on common sense and avoiding a bad potential image of Lodges or the Craft. Unlike best practices, standards address basic usability by visitors rather than site management or organizational considerations. Standards cover four areas.

Informational – Information without which a website serves little purpose to the average or uninitiated visitor. Clear identification of the site name and purpose, and working contact information that is current.

Functional – Reasonable navigation throughout the site to find all basic content with no dead internal links or empty pages, and without undue page download times. Functions such as forms must work.

Visual – Readability in terms of fonts, image and text size, colors, and general rendering of images.

Accessibility – Reasonable usability of at least basic site content by mobile devices and text-only browsers. The site must be readable on most common browsers, such as recent versions of Chrome, Internet Explorer, Firefox, Safari, etc.. Most website design and development tools create pages that fulfill this requirement. Ones hand-coded or by use of raw page-coding tools can meet these criteria by either simplified coding or a high level of coding proficiency.

The differences in standards between district and Lodge

sites are mostly if not entirely informational. This will be covered in the relevant section later in this book.

Best practices are suggestions for site development and organization for ease of use and management. Unlike standards, which address basic content and design, best practices are approaches to how the purpose and features of a website are achieved.

Best practices are not requirements, but things that are planned and done to keep content as current as possible with as few communications necessary to add or change information, as well as avoid long-term interruptions in continuity.

Most of the differences in best practices between district and Lodge sites are based on their role in communications, and further dependent upon the level of activity in a Lodge or district.

One major consideration is **Platform** — what software or system or service you will use to build and manage the site. The website should be developed such that it can be managed by multiple non-technical people (using a CMS, such as WordPress). Ideally, it should be a common open-source platform, rather than proprietary or obscure software. This means it would be much easier to train or find others to take over management when necessary.

Another major goal should be **Continuity** of administration. The site, right down to hosting, should be treated as much as possible as the property of the Lodge, district, or whatever body it represents. This is covered in much detail in the chapter "Keeping the Keys; Passing the Torch".

OUTDATED DESIGN PRACTICES TO AVOID

As of the writing of this book, the web is over 25 years old. Many of us remember how websites looked back in its early days. Some of us made websites in the 1990s, back when having a website was enough, so long as it was (often barely) readable and functional. And to some of us, old-school websites still look "normal" even though the technology, formats, and public expectations evolved and moved on.

To new web users, such sites are more than unprofessional, but damaging or even embarrassing. It may work for a "home page" for a personal hobby, but certainly not for a time-honored institution. Below are characteristics of outdated websites, sins of commission and omission. There may be exceptions to these "rules" but the list will make it obvious if a site falls into the category of "outdated" versus "modern".

GENERAL FORM

"Free" Hosting – Having your own domain name was once a big deal. Now it is expected. However, you can get away with a subdomain, so long as it's not showing off someone else's business. If your Lodge URL is "MyLodge.SomeDistrict.Org", that's not a problem, so long as the Lodge belongs to the district whose web domain you are piggy-backing on. "SomeDistrict.Org/MyLodge" is also acceptable. However, "MyLodge.KensPersonalPage.Com"   or "MyLodge.KensPublicWebsiteServices.Com" is not. The worst case is using "free" hosting that has advertising on *your* site out of *your* control. Sites like "GeoCities" (now defunct) and "Tripod" should never be used. The very sight of the address makes modern web users cringe and causal surfers laugh.

Frames – They aren't always evil. Inline frames are commonly used for things like Google Calendar, and this is indexed properly as content on the site. But frames in general can cause indexing problems and problems with mobile devices and text-only browsers (such as those used by the visually impaired). On the front end, frames should never be visible or re-sizable, to at least mask them from appearing as frames versus modern means of layout. Another error commonly caused by using frames is opening new websites within a frame, since it appears in the address bar to be under the auspices of the original site.

Old Coding – Before CSS and modern CMSs, sites used to be created primarily in HTML, with a peppering of CSS, and required a LOT of work to maintain. Only someone who knew code or had the right WYSIWYG program could publish or update content. In theory, you can still get away with this. However, it makes a transition in the future and sharing of responsibility difficult with only one person with credentials. And there is no immediate contingency in the event they are unable to respond to requests for any reason.

Disparate Design – A page made up of just a Google Calendar or forward to a Facebook page is not a website. An image of content (like a poster) is not a web page. A page that links to a handful of files or pages that are made from Word or other processed documents is barely a website. The main issue here is the consistency of presentation. The "raw" design of web pages in the mid-1990s got away with this because there were few, if any expectations and little to compare to

— it was made by geeks focusing on content alone. Without consistency, the site looks like it was made by many independent people rather than a cohesive team. More specifically, consistency means the formatting of pages – layout, fonts, colors, navigation – is expected to be consistent on all pages with rare exceptions. Each page should never have its own "theme" for its own sake.

LAYOUT

Size – When only geeks made websites, they usually had screens bigger than the average person. If they didn't test how it looked in smaller resolutions, it showed. For a while, smaller resolutions fell away, but are now back with a vengeance thanks to the prevalence of mobile devices. The general rule is that if people have to scroll sideways, ever, you did something wrong.

Navigation – The only thing worse than dead-end pages are pages where the only navigation is a link back to the homepage. All pages should have links, in the same place and appearance (somewhere "above the fold"), to all the main pages of the site. A single page linking to a bunch of stand-alone pages that do not link to each other is not a website.

Tables – These should ideally be used (if not done by other means) to separate columns of information. If used for layout, they should be invisible, specifying a border width of zero (0). Pages with lots of visible grid lines, except for a calendar or chart, are a thing of the past.

CONTENT

Splash Pages / Entrance Pages – If done tastefully, having a bold graphic or even video can in SOME cases be a nice way to first see a website. However, it should not be for its own sake, or appear to be "clever" — it's usually reserved for sites that aren't informational but entertaining. For a district or Lodge site, there is little if anything that can't be put on the "main" page of a site rather than a separate "home" page prefacing it. Again, exceptions can be made if done deliberately and with clarity of purpose to the user.

Animated Clip Art – Clip art, in general, should be used sparingly, and ONLY to accent content, or as a "logo" (such as a clip art Square & Compasses within the heading of a page). Animated clip art used to liven up a page; now it is a distraction, a cliche. Ideally, it should always be avoided, but ONE image on a page can be reasonable if it's used as an accent and not clownish.

Music – Refrain from audio unless it is embedded on a page in such a way the user can turn it on and off themselves, and it should be for a specific reason rather than background sound. Midi music in particular is the height of annoyance for almost anyone.

Counters, Guestbooks, Monitor Tests, Browser Recommendations – These are all archaic practices that should be avoided. They were interesting or clever in the days when the Internet was brand new to people, but have no place in a professional-looking site. And when a site says it is made using "HTML" and "CSS", that's like putting a sticker on a car that says it's made with "metal" and "welding" – it's pointless to people who know and confusing to visitors that don't.

It's hard to believe in 2020 some of these things still exist. And it's somewhere between awkward and impossible to tell someone who exerted so much effort as a volunteer that they gave your organization a wide tie and bell-bottoms to wear. I wish I could help the reader in dealing with such a situation, but it's not my place. But if it was my district or Lodge with an embarrassingly outmoded site, I would be furious. And if we paid for it, I would garner support to sue them.

Why so harsh, you might ask? The main argument I've heard by those who build or aren't offended by such sites is that the website is "just for us older brothers" and "we just use it for the information and don't need it to look good." What flies in the face of that is the fact that websites are inescapably an image the public will see and judge us on. We are not only speaking to ourselves, as much as we want to believe that, and we are to some extent pushing away the future in the process.

SEARCH ENGINE OPTIMIZATION (SEO)

If the web was still all about individual websites, and we were selling a product with fierce competition, a whole chapter would be devoted to this. A fraternal organization's website does not have such a need, but still should be findable, so here are the basics.

Register your site with at least Google. And yes, you can "google" how to do that. If you have a web developer build it, they should add it under their Webmaster Tools / Search Console and Google Analytics Dashboard to keep an eye on visitor statistics.

Text is king. Use more text than images, as the former is better indexed. Pictures (without 'alternate text tags') are virtually invisible to not only search engines, but people who use text-only browsers such as JAWS for the visually impaired. Within the text, use headings (with 'heading tags' if you know what that means) to set off sections from each other.

Apart from 'title' and 'description' tags (if you know how to set these), don't worry about keywords. Just make sure the words you would want someone to find you under are in the text. Above all else, be sure to have the name of the organization and its location as real text and not just a picture or text. In the end, search engines are trying to read your pages like a human — if you write your copy for human beings rather than computer algorithms, you can't go wrong, because you are aiming at the same target. Playing search engine games to "trick" them into liking you can even get you blacklisted.

Longs story short: Less is more. Don't lose sleep over it.

MOBILE DEVICE CONSIDERATIONS

If your site isn't mobile-friendly, you're not even half in the game. Phones and tablets are the most common devices right now for not just surfing the Web, but watching television. A "mobile-responsive" site is one that arranges it's content and resizes its images to fit a smaller screen. This means two or more columns become one and images don't ever "bleed" off the side of the screen. You won't' have to blow up text to make it readable, or scroll sideways.

Closely related to the accessibility standards mentioned above, mobile-responsiveness either happens by really simplistic design done a certain way, or by extensive coding for more complex layouts. Fortunately, most platforms you build a site on will hand you this capability as you go. Just be sure to test it.

CHAPTER 9

KEEPING THE KEYS; PASSING THE TORCH

There is a long-term challenge that I would be remiss if I did not devote a chapter to it. The continuity of administrative activities requires planning for those times when there is a change in who are leaders and team members. If a Secretary gets called to serve the Celestial Lodge unexpectedly, do the records in his care get tossed in a dumpster, and Lodge artifacts sold at a garage sale by his unknowing relatives? There is no telling how many times this has happened. Bank accounts and even the deeds of buildings have been discovered to be in the name of Trustees long since passed.

Technology poses more of a problem because things like websites and official email accounts are commonly forgotten digital assets. When the person handling such things leaves, who has access? I hope you can see the problem here.

But just as technology gives us additional worries in this regard, it also provides a solution. Just as a building can have a key, and bank accounts can have individuals on record who have access, so can digital assets. Passwords or access can be passed forward or distributed as necessary, so long as there is a safeguard in record-keeping.

This is the process I preach, based on two concepts — **CURB** and **DAR**s.

C.U.R.B. PROTOCOL

"CURB" stands for "Contingency Upon Retirement or Bereavement", a more polite version of what a friend in the fraternity described as, "Clean Underwear, Rogue Bus."

It is often difficult, or nearly impossible, to recursively obtain the information necessary to recover control of a web site or other digital asset in the event a webmaster or other person is unable to fulfill their duties for any reason. The best way is for more than one person to have access, or the ability to gain access in the course of an emergency or necessary continuation of control.

Such circumstances in which a CURB protocol would be used include

- Change in Personnel
- Unexpected Death
- Relocation / Lost Contact
- Non-responsiveness
- Sabotage or Asset Hostage
- Alien Abduction

Whatever the cause, the confusion and frustration that follows can be pre-empted.

Step zero is to not have anyone set up or control a digital asset unless it is clear they understand and respect the fact it belongs to the organization, not them. Many times a member will set up a website – their creation – and think it's alright to take their ball and leave if they don't want to play anymore (or get kicked off the team). The design and even text copy may be theirs, but the Lodge is expecting to use it – especially their domain name – as their own. If it is done under the auspices of a Lodge, its ownership must be made explicit.

The big problem with losing control of a domain name, website, or social media channel is that they can linger on forever. If you throw your hands up and get a new website or channel, the now "clandestine" one will compete with it, confusing people's search for the "official" site or channel. In worst cases, there's nothing that can be done at all without lawyers and luck.

But wherever you are in the process of developing an online presence (or using digital services to communicate and collaborate), you should know what assets you have, who has access to what, and a copy of the login somewhere on file.

If you can set up individual credentials with only necessary permissions, even better. Keep the main login for the organization and change who has additional access as needed. You may also be able to track who made what changes, providing transparency and accountability of work.

DIGITAL ASSET RECORD (DAR)

One way to plan for a CURB situation is to have a Digital Asset Record (DAR), meaning some file in physical and/or digital form that outlines all access links and logins for all digital channels and services. These may include:

- Domain name registration account login info
- Hosting provider and account login info
- Server login (hosting control panel, FTP, etc.)
- Website login (if using a CMS)
- ListServ credentials (if DSCNY Member) login info
- Facebook page and other social media account login info
- Google account (email, calendar(s), cloud file storage) login info
- PhoneVite credentials

At the risk of a shameless plug, I will say here that I set up the **Masonic Digital Trust** (MasonicDigitalTrust.Org) in part to provide a third-party service to store such records. If there is a gross discontinuity of leadership, I can and will investigate to determine lawful authority to turn over the record. Of course, I have *my* CURB contingency in place if something happens to me.

In summary, a Lodge or body owns and therefore has the rights to access and control any digital asset related to it. It should treat them like any other asset, and take steps to make sure they don't lose access and control under any circumstance.

ORGANIZING THE ORGANIZATION

CHAPTER 10

DIGITAL LEADERSHIP

This chapter deals with choosing people to do work and manage digital assets, as well as a proposed model for coordinating efforts across any organization made up of local chapters (Lodges), districts or regions, and a leading body.

It is based on people taking responsibility rather than having particular skills. You could call such a position a "communications coordinator" or whatever you want. Here I call it "Webmaster." It could be the PR officer, if there is one. It could even be the Secretary, who is responsible for much communication to begin with, but their comfort with a broader scope involving technology, or willingness to take on more work, may not be there.

"WEBMASTER" VS. "WEBMASTER"

The term webmaster ain't what it used to be. Email, social media, smartphones, the cloud ... Today, there are many digital tools and channels, all of which have benefit and many which can't be ignored. With so much of our lives in the virtual realm, every organization needs someone to play CTO (Chief Technology Officer). But we can't relegate tech stuff to "the computer guy" and pretend you're all set. Leadership and teamwork demand an active role in the communications and promotion of a Lodge or chapter. It's everyone's business.

A webmaster was traditionally the person who built and ran a website, top to bottom. This is no longer true, as tools have become available to allow non-technical people to manage the content of a site after it is in place. Today, a webmaster to most people means the person who built the site and handles technical details, while one or more people within a company or organization maintaining it on the front end.

An alternate way to handle this is to recognize a webmaster (small "w") as an appointed task to create and/or maintain a web site, whereas a Webmaster (capital "W") is an appointed (or possibly elected) position with specific responsibilities with regards to *using* digital communications. The former should be qualified on technical credentials, where the latter should be based on leadership, communication skills, and reliability. A Webmaster needs only intermediate digital literacy, as they can appoint or assign (or outsource) work to the former if things get too technical.

In other words, for any given Lodge, district, or other entity, the person who does the tech stuff and the person who drives that vehicle does not have to be the same.

In the Craft, the Webmaster is usually the person who isn't given any information but gets blamed when anything related to the Internet doesn't get done. Seriously though, the role of the webmaster is undefined because of the changing nature of the web and the lack of digital literacy in the general membership. A webmaster is neither elected nor appointed officially, but a Brother (usually) who volunteers or is "voluntold" to take care of it, not being clear what exactly "it" is.

At the very least, a Webmaster's responsibility is most closely tied to one or more websites used by a particular Lodge, district, or body. They may or may not develop it themselves, but see to it that it is developed and maintained in every aspect, technical and personal. Their duties may include:

- Maintaining a record of administrative logins (domain registration, hosting, CMS, etc.)

- Ensuring someone in the leadership of the group has access to this record, or a "second" for contingencies if something happens to them

- Handling technical issues or being the liaison to whatever person(s) or company handles it, or putting in support tickets when there is a problem

- Training others to use the CMS for adding and managing content they do not handle themselves

EXTENDED ROLE

As the website is the cornerstone of a web presence, a webmaster should have some relationship with anyone and everyone who handles social media or other digital assets (cloud, calendars, email account, list-serv, etc.). Such things are often integrated into a site, or the site content can be pushed out to these channels automatically. Therefore, a webmaster should expect cooperation from such individuals. But it works both ways — the webmaster should support the efforts of others in these areas and give guidance and training if possible.

Remember, to the average person, whatever happens even vaguely web-related will be associated with whoever is given the appellation "webmaster" — they may as well have their fingers in those pies.

CHOOSING A WEBMASTER

It is most important to realize a Webmaster doesn't have to be a technical developer. The best "coders" – especially the used-to-be engineer types – may be of much use but should probably stay away from managing the project overall. Beware the "DOSosaur". The same goes for the other end of the spectrum — the artist types who tend to go to no end to make everything look amazing but miss the common sense of "form follows function".

The point is that you probably have many talented Brothers to pull this off, but keep in mind very few who lay stones can build cathedrals. Look for someone who works well with diverse people, is devoted to being responsive (especially if the wages are limited to corn/wine/oil), and just has plain old common sense in communicating.

Often a webmaster may already be assigned that has solely technical skills and would not be trusted for whatever reason with the overall digital assets of the group. In this case, it may be confusing to have someone else be considered the Webmaster with the larger responsibilities while the guy doing "all the work" feeling slighted. The solution is one of diplomacy and tact, and might involve the appearance of having "co-webmasters". However, there should always be a primary person in terms of ultimate responsibility and point of contact.

LODGE AND DISTRICT WEBMASTERS

A Lodge Webmaster is the person who can demand and be given all necessary information to keep a website's content current, regardless of whoever enters the information. They may or may not be a "digital trustee" with full administrative access, but should know who is at the very least. Ideally, a webmaster should aspire to take responsibility for the calendar, either doing it themselves or making sure the Secretary or other specific individual(s) do it. They should keep abreast of trends and technologies, such as the use of automated phone trees, cloud repositories of downloadable files, social media, etc., and be willing to advise on such manners or bring in someone who can.

The full potential of a District Webmaster is much greater, as it places them in the perfect position as a channel of two-way communication between Lodges and Grand Lodge (but only in regards to digital media issues). A full Webmaster takes on many responsibilities, any of which they may delegate but are responsible for. They include:

- Maintaining a reasonably professional district website that meets basic standards and striving for best practices

- Receiving communications and guidance from Grand Lodge and provide Grand Lodge with information on district and Lodges as necessary

- Changes in meeting locations, closed and merged Lodges, and other information as may affect Grand Lodge website listings

- Local Masonic links

- Maintaining an accurate list of local webmasters and their contact information

- Being aware of, and able to access, all digital assets, including social media accounts and newsletters (even if responsibility for the asset is delegated)

- Providing resources as necessary and available to have Lodges maintain viable web presences

- Providing guidance and training to local webmasters

The concept of a Grand Lodge Webmaster is covered under the chapter, "Grand Bodies".

OFFICIAL POSITIONS AND THE WEBMASTER NETWORK MODEL

It is my opinion that an official, recognized "Webmaster" position – under any name – is vital to the future of digital communications, and therefore the Craft. It implies much larger duties than building or even overseeing. The general duties are to ensure the provision of the tools necessary for communication within and without the boundaries of their purview (i.e., Lodge, district, administrative region), as well as the necessary education to use digital media and its various tools.

In Symbolic Lodge Masonry (and similarly organized societies), the key is the districts, as they can aggregate local information for Grand Lodge and disseminate the state's collective resources and requirements to the Lodges. This ensures an ideal balance of consistency, autonomy, and shared responsibility. In practice, this means that a Grand Lodge Webmaster works with District Webmasters who shephard Webmasters in their respective Lodges (and local bodies when appropriate).

To those knowledgable about a Grand Lodge's administrative structure this should be familiar. Grand Lodges that are large enough to warrant districts have representatives of the Grand Master (District Deputies) in such districts, who in turn work directly with the Masters of each Lodge. Sure, much information is sent directly to Lodges through their secretaries from Grand Lodge, but the lines of communication for management purposes is this three-tiered model.

The model proposed here is similar. Lodges, districts, and Grand Lodge each have a position – a point man – that connects the leadership at different levels. But it is not one of authority. The Webmaster network is overlaid onto this authority. Lodge Webmasters answer to their Masters, District Webmasters answer to district leadership (or specifically the DDGM if so understood), and the Grand Lodge Webmaster would of course answer to the Grand Master.

It would be the responsibility and prerogative for Lodges and districts to appoint such Webmasters and answer for their work the same way they would for anything else that happens under their purview. But for this to work best, the whole jurisdiction would need to make this a clear expectation with the usual accountability.

One of the main problems for Webmasters is that they are relegated to the sidelines as if their work was independent instead of supportive of the many functions of a Lodge or body. One of the ways I worked to fix this is to establish an official "GLNY Webmaster Pin" — in silver color for Lodge Webmasters and gold color for District Webmasters.

And it has been my insistence that they be presented by leadership publicly or in front of the Lodge, preferably at installation, an awards night, or some other event. I have been so adamant on this point that I refused to mail a pin to an out-of-state Brother serving as a Webmaster for their district in New York. Instead, it was mailed through the respective Grand Secretary's offices to a Lodge local to him where it was presented by their Master, and a photo published for his district to see.

The following suggested language is not Standard Work in the course of installation but can be done at any time deemed appropriate.

Will Lodge Webmasters please rise?

My Brothers, you have been appointed webmaster of your respective Lodges.

Your duty is to act as trustee for the digital assets of your Lodge, seeing that the related work in communications if fulfilled by yourself or by proxy, answering to the Worshipful Master and Secretary of your Lodge, with dutiful consideration at all times for the harmony and reputation of the Craft.

You also agree to receive guidance and instruction from your district webmaster as to the expectations of Grand Lodge, report as requested the status of your Lodge's public digital assets, educate and keep up on the field of digital media as you are able and is pertinent to your work.

Lastly, you would be wise to prudently grant others the knowledge and access to take upon your duties in your absence or the succession of your office, expected or unexpected, to preserve the continuity of the work.

Will District Webmasters please rise?

My brothers, you have been appointed webmaster of a district in our Grand Jurisdiction.

Your duty is to act as trustee of the digital assets of your district, seeing that the related work in communications if fulfilled by yourself or by proxy, answering to the District Deputy Grand Master and recognized leadership of your district, with dutiful consideration at all times for the harmony and reputation of the Craft.

Lastly, you would be wise to prudently grant others the knowledge and access to take upon your duties in your absence or the succession of your office, expected or unexpected, to preserve the continuity of the work.

{To all the above Webmasters}

You have been hereby presented with a webmaster pin from the Grand Lodge of the State of New York. Wear it as a badge to distinguish you among your brethren as having been conferred the duties thereof, with the right to request and receive the materials and tools necessary for your labors.

A lapel pin or title may seem trivial. The point is that someone doing this work, especially in the manner prescribed above, is ideally a visible and active part of every Lodge's and district's leadership team.

CHAPTER 11

LODGES AND CHAPTERS

L odges and other local, subordinate bodies are the heart of a jurisdiction, and Masonry itself. Each has its own unique identity, culture, and place in the community. This is either reflected online, or it is not. If a Lodge doesn't have the resources to have an online presence, they should make every effort to take advantage of district-wide channels to promote their meetings, events, and programs. But ideally, managing their own web presence gives them control over such things.

LODGE WEBSITES

Lodge websites should clearly identify where they are, being geographically clear enough to be understood by people not from the area. (Remember, you may not intend the site for non-members, but it will be linked to by the Grand Lodge website, district website, and search engines.) The general public is searching for information on Masonry and will find your site(s).

Necessary informational elements (standards):

- Lodge name and number
- Physical Address

- Meeting times
- A current and active contact phone number and email.
- Prominent link to your Grand Lodge
- Link to official district website
- A calendar or event list, or links to such list(s) as available. It may use the district calendar as a substitute.

LODGE WEBSITE BEST PRACTICES

The purpose of a Lodge website is to provide information on the Craft with a specific emphasis on the information regarding the working of the particular Lodge, i.e., as it relates to the community it serves and exists in. It should, of course, also provide relevant information for the members of that Lodge.

If the Lodge owns the property in which it does its work, rental information, if applicable, should be given with a way to contact the appropriate person to handle inquiries.

Other helpful content:

- Interactive map
- Interactive contact form
- Downloadable, printable membership petition (PDF)
- Typical Additional content:
- Elected and appointed officers

- Charity and other projects associated with the Lodge, including Lodge scholarships
- Links to concordant bodies it has a close relationship with

SOCIAL MEDIA

My recommendation for most membership-oriented organizations is to have a public Facebook Page and a private, members-only Facebook Group. Many things will be posted on both, but the latter allows for more freedom of discussion, and notices of strictly internal affairs.

For social media in general, Lodge accounts should follow or like the accounts of other local Lodges and bodies, as well as that of their district and Grand Lodge.

EMAIL, ETC.

Like paper mailings, district leadership and other local Lodges and bodies should receive general digital newsletters and notices. Be particularly mindful to send such things to those running district communication channels, i.e., a District Webmaster if there is one.

CHAPTER 12

DISTRICTS AND REGIONS

L odges and other local, subordinate bodies do not each exist as an island. They should work together, and that should be reflected online. The district's web presence should be the most useful means.

DISTRICT WEBSITE

In my jurisdiction (The Grand Lodge of the State of New York), all districts are required to have a website (or share a website with another district) that meets or exceeds the standards explained previously. Lodges must have a website presence, either by having their own site, or a page provided for their basic information on the district's site.

District websites should clearly identify the districts(s) covered by the site, as well as geographic region for uninitiated visitors. (Remember, you may not intend the site for non-Masons, but it will be linked to by the Grand Lodge website and search engines.) The General Public is searching for information on Masonry and will find your site(s).

STANDARDS

Necessary informational elements:

- Prominent link to their Grand Lodge
- Listing of Lodges within the district(s) covered, containing current basic information

  - Lodge name and number
  - Physical Address
  - Meeting times
  - Link to the official website. In absence of such link, a contact phone number and email address (with the current name of secretary preferred)

- A calendar or event list, or links to such list(s) as available
- A list of the current district leaders, and a means of contact for at least one key person

BEST PRACTICES

The main purposes of the district websites are to provide local information on the Craft. However, equally important is how it looks and works for the general public, as it is public, and as such should also consider and allow for those seeking more information to receive it.

Ordinarily, there should be sections of a page or even separate pages (or links) for

- Lodge listings

- Local Lodge of Research and School of Instruction (if applicable)
- Concordant and appendant bodies with logos and links
- Events (See chapter on Calendars and Events)
- Local Lodges in amity, such as Prince Hall Affiliation

It is ideal to have links to various projects and programs, including but not limited to

- State and local programs, such as the Masonic Safety ID Program
- State-level charity (such as The Brotherhood Fund in New York)
- State-level and district scholarships
- District-wide events
- Other district-hosted content

If any of the above does not have an official site, a link can be given to a page on the site itself devoted specifically to the Lodge, body, or project. This is particularly helpful for such entities that have are too limited resources and membership to justify having their own site.

SOCIAL MEDIA

The importance of social media channels depends on how many programs and events are held on the district level. At the very least, they should support the accounts and websites of the Lodges and bodies in their district.

DISTRICT WEBMASTER & DISTRICT DEPUTY GRAND MASTER (DDGM)

The following is a general outline of the relationship between a proposed District Webmaster and the District Deputy Grand Master (or its equivalent in other organizations). I must make clear once again this is for my jurisdiction and is not official protocol, but is what we have been providing as counsel for leadership and members.

Who owns a district website?

The short answer: The district. The long answer: It belongs to the Lodges of the district, under the jurisdiction of Grand Lodge, and is run by those who run the affairs of the district. This could be a Purple Association, Lodge Council, or simply the Grand Lodge Officers (DDGM, SO, SB, etc.). No person owns it any more than a person – even one sitting in the East – owns a Lodge building.

Who owns the intellectual property of the site?

A web designer has the copyright of the intellectual property of the design itself, but it should be made clear beforehand they are giving it to be used freely for its intended purpose and has no other claim except credit if requested and agreed.

The actual files and located somewhere at a host's computer (webserver) and are due rent. That means the district does not own the machine physically, part of which is the live copy of your site, but have the right to use it for the live copy of those files as long as you have an arrangement.

The content of the site may be written by many people, even regular contributors, such as the AGL or DDGM, and the rule is the same. People have a right to be credited for their writing, but we would hope they give explicit or implicit permission of unlimited appropriate use from within the craft. The same goes for photographs and other media.

What is the DDGM's role?

They have no direct managerial responsibility unless they so choose, but are responsible in the same way they would be responsible for any other communications within their district.

To clarify: It is not the DDGM's job, or any Grand Lodge Officer's, to babysit a website. Apart from contributing articles as they are willing, it should NOT be their job. However, the DDGM is responsible for the progress of their district and all that entails. They should know who the communications people are, work with them, and find new ones as necessary. The website is part of that. The webmaster(s) should be accountable and responsive to the requests of the DDGM, and the DDGM should avoid micromanaging or second-guessing a competent webmaster.

The trustees of a web site and other digital assets – those that ultimately have direct control of it – fulfill a long-term role. Webmasters and others who have the keys need to be part of the immediate team, but also serve as a stable continuation and consistency of the district over many Grand Lodge Officer terms. This, as in many things Masonic, is where leadership and teamwork are more useful than authority. And if done well, takes the burden off the DDGM.

What if Lodges do not wish to participate?

A district website is required to have useful basic information on all Lodges in the district, or links to their own websites which should have such information. A calendar is another thing not to be taken as optional. If a Lodge does not cooperate, that is a leadership issue that must be addressed by the DDGM.

The bottom line: All Lodges need to be represented usefully online either by themselves or by the district if they are unable or unwilling. For some small (or failing) Lodges it doesn't make sense to have their own building. But if they exist, a web presence of some kind – even a page on a district site – isn't a luxury. Grand Lodge isn't going to send the cavalry into the quarries to fix this, but it will be done. No Lodge is an island, and the success of your district depends on their cooperation in this matter as well as all the usual ones.

Who handles the costs of having a website?

There are usually three expenses to cover: the domain name; hosting; and web design. Some of the services many be donated, or the expense covered by a Brother, but there is no rule except that financial obligations are met to ensure the development and continuity of necessary services. This should not be a significant cost, and if necessary, and upon agreement of Lodges and district leadership, can be easily spread across Lodges or even other Masonic bodies wishing to derive benefit.

Reasonable costs at the time of this printing: $15-$25/year domain name; $100-$360/year hosting; $750-$2500 one-time website development

What about Concordant and Appendant bodies?

In the strictest sense, there is no requirement to even

mention them. However, regardless of differences of opinion as to the concept of the "Masonic Family", it would be prudent to follow the example of Grand Lodge's website and be sure to at least link to information regarding such bodies that have a presence in your district geographically.

A district may also wish to reflect the relationship cherished by the Grand Master and so many others by allowing them to contribute content (news, integrating calendars, etc.). And if the people setting up and administering the site wish to provide pages for Lodges and chapters who do not have their own website, that is a great way to fill in the empty space in the overall Craft web presence for your area.

Are we allowed to mention Prince Hall Masonry (in the Grand Lodge of New York)?

Recognized Lodges under the jurisdiction of the Most Worshipful Prince Hall Grand Lodge of the State of New York are absolutely welcome to be mentioned. To what extent is our obligation? I would follow the example of Grand Lodge and list them. Our relationship is outlined at NYMasons.Org and an official listing of PHA Lodges in the state can also be found there as well.

However, there are reasonable limitations. We should not appear to represent the Lodges of another jurisdiction, only in matters of mutual promotion, benefit, and joint activities. Having their calendars integrated into ours is a grey area. I am working more and more with Lodges and districts within their Jurisdiction to ensure they can represent themselves fully and independently online. Just be sure this is understsood as a matter of respect and sovereignty to each institution, not segregation between Brothers.

ADMINISTRATION

The long-term plan is to have at least two people in the district have the "keys" to the site, domain, and hosting, just as the trustees of a Lodge would for a building. And this applies to all digital assets, not just the website.

These people need not have technical skills, but access in the case of an emergency or continuance in the event a webmaster moves, resigns, enters the Celestial Lodge, etc.. The webmaster(s) should have reasonable documentation available for smooth succession.

CHAPTER 13

GRAND BODIES

E ach Masonic jurisdiction and grand body in Masony (and other similar societies) is unique and there is no easy answer on how to best administrate with regards to Digital Media. But I will share my thoughts based on my knowledge and experience of the Grand Lodge and bodies I am familiar with.

On one hand, it is an entity onto its own, and its web presence should reflect that, even if it gives credence to and information on subordinate chapters and bodies. It probably has its jurisdiction-wide programs. If it is a not-for-profit, activities related to that must be prominent in all its publications.

SUBORDINATE LODGES

A Grand Lodge should in some way list or point to its Lodges. A Lodge locator (map) and a directory is ideal, but not always feasible if the jurisdiction is huge with constant changes. A Grand Lodge may or may not be able to plug into its membership database, which would likely keep track of these and feed them directly to a website.

The way I've handled it in New York was to have a Lodge locator with basic contact information and the times they meet. This is good for visiting Brethren or people looking to join a Lodge. However, districts are an internal, administrative distinction, so I stopped posting that list with its constituent Lodges. In the future, I may simply link to all the district websites in the hope they are diligent in accurately maintaining their local information.

A GRAND WEBMASTER?

A "Grand Webmaster" is actually a thing in rare cases, with its own Emblem and Apron. I'm not sure that the title fits what I do. I am the Webmaster (lower-case and capital) for the main public website for my jurisdiction, NYMasons.Org. But there are many other websites for the Trustees' projects, Grand Lodge programs, and social media channels I have nothing to do with. I am simply a member of the Communications Committee and work with the Technology Committee.

It makes sense that our Masonic Medical Research Institute, the Chancellor Robert R. Livingston Masonic Library, the Masonic Care Community, and some others have independent sites. They are entities onto themselves, and it would be too much work for one person to directly manage them all. It just wouldn't make sense.

Instead, I see my role as educational, showing districts and Lodges how to work together and with Grand Lodge. But if there was a "Grand Webmaster," the following is how I would envision that role.

A Grand Webmaster would have an overview of all the state-level digital assets. This would not be a matter of authority, but coordinating and building teamwork. It might mean development and education regarding standards and best practices. And it would mean shepherding disparate social media channels and internal communication methods (as it is all too easy for every chairman to create their own "Grand Lodge" channel that competes and confuses with the others).

The challenge is that this is about technology but not *for* technology. When telephones became a thing, I would be surprised if companies and organizations created a "telephone committee". Instead, it became a part of the process for overall communications. So should it be with the Internet (and cellphones, etc.)? The tail should not wag the dog.

This goes back to who you choose to be "Webmaster", the tech whiz or the person responsible for how it integrates into real-life goals and processes. Maybe it needs to be a little bit of both. (I would rather be a businessman with IT skills than the other way around, and I've built a living on just that.)

OPPORTUNITIES

But one thing is sure: any grand body is in a position to lead the way. It can get caught flat-footed when Lodges do new things the "higher-ups" haven't planned for or even understand. It can respond with over-cautious edicts, or fail to respond when a line is crossed. Or it can harness those in the Craft who are doing this work, letting such Brothers travel and teach others to lead the way on behalf of the whole jurisdiction.

When I travel and present Webmaster and Communication Conferences around the state, I try to back away a bit and let locals do most of the talking. This makes it clear that we have resources in our own backyards that we need to start tapping into. I truly believe a Grand Lodge can facilitate this from the bottom up, and to great effect.

CHAPTER 14

CALENDARS AND EVENTS

{Note: There are a lot of mechanics in this chapter that could become inaccurate when the technology and user interfaces change. The principles will still apply.}

Today, people demand access to current and upcoming events of the organizations they belong to, any time, anywhere. This doesn't mean a list of events on a website anymore, and certainly not a printed grid or list. Only a digital calendar can be seamlessly shared, imported, or combined with other calendars. There is no need for sending them out or notifying anyone of changes — they are as current as the devotion of those whose responsibility it is to keep them accurate. Accommodating those without the ability to communicate through modern channels should be treated as the only necessary exception. And today, it is easier than ever for the average person to set up such a calendar.

But I've placed this chapter on calendars in this section for a reason. It's not because it spans everything we've touched on to this point – communication, public relations, events, and administrative planning – but because calendars are the biggest pain point in terms of organizational conflicts, and it can only be resolved by coordination at all levels.

The need to resolve calendar conflicts was the impetus for a Masonic town hall meeting that resulted in my occupational involvement in Grand Lodge. Events were constantly scheduled at the same time for various Lodges and concordant and appendant bodies, even though some were announced months in advance. This had to end, but there was no one calendar, except for a Masonic newspaper that came out only a few times a year, and not every event was reported to the publisher.

Since then, I put in place solutions that have worked in my own district and in other parts of the state. Some other districts may have had other paths to success. Some haven't addressed it at all, or it isn't as much of a concern because there are fewer Masonic activities.

The point of all this is that if you do this, there must be near-one-hundred-percent participation. If it is centralized, there will need to be constant communication between the Lodges and bodies and the one maintaining the calendar. It's a lot of work for one, and doesn't solve the same problem that happened with print methods — people don't consistently go out of their way to share their information.

So I created a model where the responsibility was distributed directly out to the individual Lodges and bodies, who could both use their calendar for their members and website, but it would be aggregated into a master calendar for the district. This maintained the Masonic ideal of Lodge self-management while fostering teamwork in the larger Masonic community.

To be honest, it takes a little re-education and encouragement to keep the flywheel spinning. Sometimes a

Lodge trying out some shiny new alternative system disrupts the flow if they didn't also keep up the one everyone else was on board with. Their events simply disappeared from the district calendar.

More than any other administrative effort I'd undertaken, this is the most like herding cats. But when it works, it works.

THE AGGREGATED CALENDAR MODEL

Please note that specific instructions here are given for Google Calendar, as it's the easiest and most common system. Webmasters or others wishing to use another platform should make sure it is in a standard format (ICS) that can be just as easily imported or aggregated with other calendars. If it's proprietary to where you can't view it in your regular phone app or have others add it to a regional calendar without extra work, most of the value of a standard digital calendar will be missing. Don't use those.

LODGE AND CHAPTER CALENDARS

First, the buy-in. There are three reasons for any Lodge or body to have a public digital calendar: It can be checked by members or interested parties for changes as they are made; it can be made visible to people who wouldn't otherwise know about events, such as visitors from other places; it can be seamlessly added to a regional calendar and requires no effort once set up.

In other words, a calendar isn't just for members who know what is going on by word of mouth. It is for visitors, and more importantly, being aware of potential conflicts between Lodges and bodies in any given area.

If your Lodge (or chapter, body, etc.) does not have a dedicated Google account, create one. This does not belong to the webmaster or current leadership but the organization itself — an asset (along with access credentials) to be passed along as necessary over the years. This account automatically comes with an array of products, such as email (Gmail) and cloud storage (Google Drive), but what we are concerned with here is Google Calendar. Hint: when you are logged in and are not in Google Calendar, simply click the nine dots in the upper right to choose it.

The main calendar should be set to "public" (on the settings screen for the calendar, not general settings). Also on this page, you can get the code to embed it on your website if you choose to.

If you want someone to be able to add events to the calendar without giving them access to the account itself, you can add their email (associated with their own Google account, which they can create if they don't have one) on the "Share this calendar" page. This is also useful for the webmaster to manage it from their own account instead of logging in. Hint: if you add an event while in your own account and it doesn't appear on the Lodge's calendar, you probably didn't specify which calendar it belonged to so by default it was put in your own. You can change that at any time by clicking on event details and using the drop-down setting for "Calendar".

Make sure you let people know the email address of the Google account so they can add it to their own Google calendar view, phone apps, calendar programs, etc..

IMPORTANT: Inform the district webmaster of this address so they can add it to the calendars followed by the district account. This is required for integration into a district or regional calendar.

If there are private events you want to track (such as hall rentals) using this account, you can set individual events to private, or use a private sub-calendar.

A NOTE ABOUT EVENTS

Events can be added (or updated) by anyone given such access and responsibility. It could be the Master, Secretary, or someone delegated. If you don't have details of an event yet, setting it as a general "all-day" event is helpful for internal and external coordination. But once the details are available, whatever is known and necessary should be included:

- Name of Lodge or Body – Preface the event with the Lodge name so it will be distinguishable from events from other calendars when combined with others (at the district level)

- Location – Do not assume people know where it is, such as potential guests, and a clear address will automatically translate to a phone's GPS.

- Cost, if any

- Attire – Don't make assumptions, as visitors will not know local customs

- Who is allowed to attend (i.e. Master Masons, wives, children, EAs and FCs)

- Contact information or other means to RSVP if required

Hint: If an event is canceled, DO NOT DELETE IT. Mark it as "CANCELED" in the event name so people will not assume it was not entered on the calendar or wondering what is going on.

THE VALUE OF DISTRICT CALENDARS

A district calendar is one of the single most useful things a district website can provide. The days of waiting for Trestle Boards to be mailed and data-entered as received are over. It is up to the district leadership to make this one fact clear: if a Lodge or body wants their events to appear on the regional calendar, they must properly maintain their own digital calendar. And this calendar must be made known to the district webmaster.

A webmaster should maintain a spreadsheet of Lodges and bodies that have (or should have) calendars to be integrated into the composite calendar, including a current contact for each. As many Lodges may not be on board yet with this system, hand-holding and prodding by the DDGM may be necessary. In the end, communication is key to having everyone understand what they have to do to "exist" on the public regional calendar.

HOW IT'S DONE

In the county's dedicated Google account, all the necessary calendars should be listed under "My calendars" (if they are actively shared with the account) or "Other calendars" (subscribed by adding the respective account's email address in the "Add a friend's calendar" box and hitting enter). To make a composite calendar, select any calendar's "Calendar settings" and click on the "Customize the color, size, and other options" link above the embed code. This will open a new page where you must check-box all the calendars you want to combine to generate the appropriate code. Be sure that if you choose to display the calendar name (checked box by default) that it does not have a sub-calendar's name — you can change it to however you want it to appear.

Hint: If an individual calendar's name is off-kilter (all lowercase or the email address instead of the name in plain English), you can change it under "Calendar settings" for that calendar, even if you do not own or control it. This determines how it will look on the aggregated calendar. You can add or remove Lodge numbers, or however you want them to be consistent.

ALTERNATE METHODS

A district could use their account to create sub-calendars for Lodges and bodies, doling out access or entering events themselves, but this brings up questions of who has control and who has responsibility. (And if you absolutely must do the work for a Lodge because they won't cooperate and do it themselves yet are expected to have the result, you should expect compensation, just as Grand Lodge may start charging data entry fees for Lodges that do not use MORI.)

Using other or multiple calendar platforms can be more work to set up, but it can be done. If several calendars (instead of a rare exception) are using another standard-format calendar other than Google, many plugins and other software can be used to combine such calendars. Just be aware this is an inconvenience to a webmaster who isn't already doing this. This small point of calendar standardization has a tremendous benefit collectively.

FINAL NOTE

Fraternity is about who we are and what we do, and nothing we do is in a vacuum. With rare exceptions, information about our events needs to be public, and it should never be assumed everyone who wants or needs to know *will* know. An online digital calendar trumps any other calendar because of its availability everywhere, any time. It is *de facto* the "official" one even if it is wrong or neglected, and therefore maintaining it requires a steadfast commitment.

And even if no other promotion of an event takes place, people from across the local Masonic Family will at least be informed and have the opportunity to attend on a much grander scale.

CHAPTER 15

DIGITAL SQUARE CLUBS

L ast but not least, I share with the reader an opportunity to the benefit of all. As a means to fulfill directives of Grand Lodge and assist in the proper use of technology in general, I created the concept of the "Digital Square Club".

A Square Club is an informal group of Masons who meet for fellowship and discussion. They are often themed by location or profession. Currently, I administrate the Downtown Buffalo Square Club, a monthly lunch gathering of Masons and guests who work downtown or are otherwise available at that time and place. Companies, such as Bell Aircraft, had their own Square Club for employees who were Masons back in the day. The largest one in New York currently is the Police Square Club of the City of New York.

A *Digital Square Club* is for those who serve, or wish to serve, the Fraternity through digital media — photography, video production, web design, graphic art, social media management, electronic communications, or even digital preservation and editing of historical wikis.

In 2015, I founded the "Digital Square Club of New York" as well as the "World Digital Square Club" group on LinkedIn. Admittedly, the last of these didn't gain traction and I've had little energy to nurture the former as much as I had wished.

However, leaders from these ranks have slowly arisen to help me hold several well-attended Webmaster and Communications Conferences around the state.

The purposes of a Digital Square Club can be:

- To share, discuss, and develop best practices
- To provide a talent pool for Lodges and bodies needing help
- To give recommendations to Grand Lodge regarding digital matters
- To network, collaborate, and mentor members

It need not have a budget or even an official membership list. All that is needed is a list of those interested and people willing to help organize communication and events.

I would encourage every jurisdiction to promote and accommodate the forming of such a club, and I am at your service if you seek guidance on how to proceed.

SECTION III

PUBLISHED ARTICLES

The following section contains all the full articles I have written in the Grand Lodge of New York's quarterly magazine, the *Empire State Mason*. Some have been reprinted in other jurisdictions and bodies. The choice of topic and wording often reflected the circumstances of the times or was intended to bolster the theme or initiatives of the Grand Master. Some are inspired by particular events, or to promote upcoming ones. It is my hope that the central message of each of these will be useful to future readers, regardless of where or when they are, and I encourage you to apply it to your own time and place.

Any of these may be reprinted by permission of the author and the Grand Lodge of the State of New York. No permission is needed if used as a presentation or for discussion at a Masonic gathering.

On a personal note, I don't recall how I started writing articles for the *Empire State Magazine*. I probably just submitted the first article (in 2015) and was surprised it was published. I'm still surprised I have what is now the column "From the

Webmaster". What I do – and I often admit this in my articles and presentations – is housekeeping. It's only important if it doesn't get done. I am young in the Craft, and until recently have had no special title other than that greatest-but-equal appellation of "Brother".

And it surprises me how many people around the state have recognized me by the tiny photograph. I'm not sure if all this is humbling or just swells my head, but I've tried to steward that quarterly, 500-word space well. At times, I've strayed from housekeeping themes quite a bit into the esoteric, barely touching on technology. My target is always what it all means to our benefit as individuals, the Fraternity, and the world. And in the end, I want the Square and Compasses on my memorial, not a chiseled "The Web Guy". I pray that no matter where my occupational intersection with the Fraternity takes me, I never forget what I have come here to do.

THE LODGE, THE WEB, AND THE HIVE

Our world of technological wonder may sometimes seem at odds with Masonic values, ancient and timeless. But how we practice and give voice to Masonic Truth depends on the times and tools at hand, and the ears of the world around us. Understanding a timely need to modernize our communications in this ever-changing world, Grand Master William Thomas took no delay in setting things in motion.

Deputy Grand Master Jeff Williamson shares this goal as head of the Communications Committee. He speaks widely that we should consider all the generations of Brothers in our Fraternity. This includes the use of paper, emails, websites, and social media. It is an unspoken charge of "no Mason left behind". It also means being accessible in ways that are expected for upcoming generations of Masons, ways that did not exist a few short years ago.

MW Thomas set forth a list of "Best Practices for social media", the electronic newsletter "Hiram's Highlights" came into being, and I was asked to develop a new Grand Lodge site. I was surprised and a bit intimidated. This wasn't like other projects I've done over the last 15 years. It was for something both grand and dear to me, and it has proven a wondrous opportunity to both serve the Craft and meet Masons across the state.

As a professional web developer, I do not believe in using technology for its own sake. Here is a chance to use digital tools for more noble and glorious purposes. What is the role of the Internet in communication between Grand Lodge, the districts, Lodges, and concordant bodies? How does it brand Freemasonry to the uninitiated? How can it be used for Masonic education? These are questions we aspire to answer.

Now that the new site is launched, calendars are being integrated, and documents and forms shared on the cloud, we are in full swing to ensure our communications are relevant to 21st Century demands. But it doesn't end at Grand Lodge. Many districts and Lodges have their own website; others do not. Some use calendars that can be combined to avoid local event conflicts. Many do not but could find it very useful once put in place. Some websites have little useful information, missing things as simple as contact information or a physical address. Some of them are abandoned, outdated, or unofficial, yet publicly representing the Craft for better or worse. To fix all this might be like herding cats.

The Next Step

The standing challenge for districts to each have their own website is now a working plan. The first step will be to inventory all official working district websites and link to them to and from the Grand Lodge website. The second step will be to provide the resources for all the remaining districts to have such a site. These sites will ideally be run by brothers within their respective districts, and training will be provided as necessary. The format made available is user-friendly enough for anyone with basic computer and word processing skills.

But this will not be micromanaged by anyone. This will not be herding cats.

At the St. John's Day weekend mass, RW Grand Chaplain Wainwright McKenzie spoke of the industriousness of the beehive. This caught my attention for two reasons. The first is that my daughter and I started keeping bees this year. The second is that it came to me one day that if a webmaster was an officer of the Lodge, their symbol might be a hive under a quill. Yes, I am suggesting that for communication, we can work as a hive, but what does that mean?

There is a saying in beekeeping that the beekeeper wants the bees to do what he wants, but they want to do what they want. Perhaps this is the challenge of balance in an institution such as ours, a harmony of leadership and individual conscience and action. In a hive, each bee may dance to tell the others where and how far away a resource is, such as water or forage. But each has their duty, moving over the course of their lifespan through the "chairs" of nurse, guard, forager. Their ritual is written in their DNA; their landmarks are the instincts of the survival of the hive.

But I do not think the Grand Master is a queen bee, and we merely workers and drones. That doesn't sound very Masonic. The role of leadership is perhaps more like the beekeeper. They provide the structure, support, and ever-watchful protection and care of the hive's health. Our goal is not to put the web presence of New York State Freemasonry under one controlling team, nor dictate how the job gets done in the local quarries. The plan is to make it as easy as possible for us to better communicate and have our efforts be in the same direction. We must be on the same page.

Over the upcoming months, we will be proposing and publishing a set of recommended best practices and common-sense standards for official Lodge and district websites. An online repository is being developed for use by webmasters across the state. It will contain logos, QR codes, and other graphics, as well as commonly used forms and content. There is much more to come, but this is a start.

It is no small task. It will take time. We intend not merely to lay stones over the next two years and start over yet again when the torch is passed, but, as RW Williamson puts it, "to build a great work of beauty". Our trestle board is framed with efforts that we believe will not simply fix a problem or sustain us for today, but create a framework that is bigger than any webmaster or website, a work we can all share that endures for years to come. And that, I would suggest, is very Masonic.

DIGITAL TRUSTEES

Who owns your Lodge building? This seems like a simple question. But when asked a number of years ago, many Lodges weren't sure. Some turned out to be in the names of dead trustees from years past! Here's another question: Who has a set of keys? Does anyone know for sure who has access?

This can be difficult to manage with real property – and neglected for virtual matters. The Secretary has access to MORI and may be responsible for physical mailings, but who manages other communications? Who is in charge of the website? Do you even know if your Lodge or district has one? In other words, who owns your online presence?

Another way to communicate is a phone tree. More and more organizations are replacing (or supplementing) this with automated "robo-calls" through an online service, such as PhoneVite.Com. Under which officer do such duties fall? There is a large enough divide in digital literacy that the secretary may delegate such things, if they are given the task at all. One Lodge's Master set up an account (in their name), then the next master created a new account (in their own name). Fortunately, the Master after him showed forethought and created a Lodge-owned account that could be passed on as necessary.

It's even worse with websites. It often falls to a brother

who simply knows more than the next guy about "computers". A lot of work and effort may be put into it. But then they aren't given the information they need to keep it current, or they become too busy. Or they move out of state. Or leave the Craft. Or travel to the Celestial Lodge. Maybe someone took the initiative to make a site for their Lodge but it was never officially used, and an approved site was built later, making the first a "clandestine" site still in public view, with out-of-date information. The lights were never turned off, creating confusion. How many orphan or clandestine sites are out there in your name?

Like no current member having their name on a Lodge bank account, the domain name (the 'dot-com' or 'dot-org') might be in the name of someone you cannot reach (or will not cooperate). It can be fixed, but it's a tedious process to recover such an "asset". And getting a new name leaves an old, uncontrolled one out there.

Or maybe you have control over your "name" and need to take control of the site itself. Who is paying for hosting, and will it lapse? Who has a username and password to make future or emergency changes? In other words, who has the keys?

The solution to cure and prevent all these problems is to treat a website, common email account, cloud service for storing documents, etc., as "digital assets" of the Lodge. The webmaster doesn't own the website. Or do they? This needs to be settled.

What is a "webmaster" anyway? It used to mean one person who both builds and maintains a website. That doesn't have to be the same person any more, since new sites can be managed by anyone with word processing skills. Now you can have a site built and supported by a company or technical person, while almost anyone within the Lodge takes responsibility for its content and use. (And a lot can be automated, so state, district, and local information doesn't have to be kept separately by everyone.)

Think of a "webmaster" as a "digital trustee" – someone who has the keys, responsible for control of the site, online files and accounts. Just like regular trustees, there should be more than one. Even if there's only one brother with skills and desire to do the work, there should be a clear plan of necessary information and access by at least one other person in case the first becomes unavailable.

Remember, in Masonry, nobody owns an office. That principle is one of our great contributions to Western Democracy. When a brother leaves a chair, the work must continue. But the webmaster is neither officially elected nor appointed. They just sort of happen. They are a fifth wheel, often kept out of the loop when needed the most. And when they come and go, we often are left to start from scratch and can't even clean up what already exists.

I suggest we look at this role carefully, differently. Official or not, it is a responsibility that is not at all superfluous or peripheral to the efficiency and survival of the craft. Our "digital trustees" need to be empowered and accountable. They deserve recognition and must accept direction. It is one of my goals as a member of the Communications Committee to bring this grey area into the light.

Under the direction of the DGM, RW Williamson, and with help from many brothers throughout the state, we are now actively working to catalog district sites and their webmasters. And instead of trying to establish top-down control, the plan is to guide and give resources where they are welcome. In fact, the way I hope to achieve this is by collaborating with webmasters across the state, to develop standards and best practices together. If there is interest, we may even form an "internet Square Club" where webmasters and other interested masons can meet (online and off) for fellowship, support, and sharing their expertise.

There is no need to pull everyone kicking and screaming into today's world. All we need are enough craftsmen willing to establish, digitally, the beauty and strength the craft requires here and now, enduring into the future. If you are interested in being somewhere in the roll call, contact me.

THE PLACE AND IMPORTANCE OF TECHNOLOGY

I preach the technological needs of the craft. So the following may seem a contradiction. People may find digital communication a mystery, or another task to keep an eye on. Some are excited to use new ways to do old things. Younger brothers just take it for granted as part of everyday life.

But it isn't Masonry.

More than ever, people desire to pass from the profane to the personal. Words are conveyed mouth to ear. Knowledge is stored in the faithful breast, not a hard drive or the "cloud". What better way to convey truths than physical ritual? The meaning isn't in the print of the Standard Work and Lectures. No matter how much is "revealed", you can't impart Freemasonry online. You can't email the Real Word. As great as it would be to have a phone app for prompting, or text-searchable Constitutions, the form it takes is just housekeeping.

It seems a no-brainer to turn off your phone in Lodge. Outside communication diminishes a sacred retreat of friendship and virtue. But what if it's your appointment book? Events are announced and people scramble to write it down. Why not a hand-out? Better yet, make sure it's on the website calendar. Technology ought to bring us less distraction, not more.

Consider the brethren of old, who had to take buggy and barge to attend Grand Lodge. Planes, trains, and automobiles put us in same day's travel, while officers and committees meet in virtual conferences from the comfort of their homes. We'd be foolish to go back. But we'd be foolish to think this the practice or point of Masonry.

Consider the 18th Century. The western door was narrow, not by prejudice of social class, but limited time after toil. With the industrial revolution, regular work created a middle class with concepts of "vacation" and "time off". The masses found time to participate in clubs and fraternities.

In the information age, a secretary can do in minutes what used to take hours. A rare few still prefer hours. But promises of plentiful free time fall short. We simply do more work in the same time and find other ways to distract ourselves.

Why? Because technology can change the world a man lives in, and how he lives in it, but it doesn't change the man. That's our job. That's Freemasonry.

Digital tools are more than a convenience. But wisdom is found not only in how you use a tool, but in deciding when NOT to use it. A calendar accessible by smartphone is a demand of the times. What we put on it is the measure of our work. An email or Facebook post can alert us of sickness and distress, or a brother laying down his tools. But a phone call or visit is what is good and true.

Many within the craft are working hard to handle these temporal issues of technology. We work to make it easier for the real work of Freemasonry to happen.

CROWDSOURCING THE CRAFT

Exciting things are happening in our Grand Jurisdiction, expressed eloquently by Grand Master William J. Thomas at St. John's Day weekend. Some initiatives are already prominent on NYMasons.Org, such as "Restoring Civility in Society". More will be coming in conjunction with a new, more modern and mobile-friendly website.

To achieve that, I decided to call out for photographs of Brothers sharing fellowship and doing good works, rather than using stock photos. The response was substantial. Brothers from around the state shared photos and albums, and some were exactly what we need. Many were professional quality, as many brothers are professional photographers.

We also had our first "webmasters conference" in Utica this year, teleconferenced to brothers who couldn't be there in person. We spread the word on Grand Lodge's direction, but it was a perfect opportunity to kick-start a network of Brothers to share knowledge, advise leadership, and be an ongoing talent pool for the Craft.

This "Digital Square Club" is open to any Mason who wishes to serve the Fraternity through digital media – photography, video production, web design, graphic art, or even digitally preservation and editing historical wikis. Together we collaborate on standards and best practices. Reach out to me to be part of that TEAM*.

Today this is called "crowdsourcing". But everything old is new again. Lodges raise Brothers and Grand Lodges grant charters, but it takes Brothers to form a Lodge, and Lodges to form a Grand Lodge. Funding to build the Grand Lodge Building and what is now the Masonic Care Community started with a single Brother's silver dollar. Many works can be attributed to individual Brothers, and grand undertakings by informal collections of such Brothers.

The idea that leadership is necessary but authority is derived by those governed was a cornerstone laid by certain Founding Fathers. Our nation may waver from such ideals, but our Brotherhood need not. When we recognize and harness potential in our quarries, rather than focus too strongly on those called to serve as leaders, we realize each of our right and responsibility to contribute. We choose leaders to make the hard calls, not to do all the work.

This is not really doing things a new way, but living our ideals. For example, Brothers should not wait for an invitation to contribute news to Hiram's Highlights. We surely have an untapped army of "reporters" better suited to do these things than appointed officers who are busy with other duties. Bro. Andrew Roberts, the editor, would be pleased to hear from you.

We don't have to buy stock graphics with artists sitting next to us. Together we can create a repository of royalty-free Masonic images. And there are Brothers who can do PR, or could be trained to do so – and yet that seat goes unfilled in most Lodges. I could go on, the point being that we shouldn't "outsource" but "crowdsource" those things that would be as simple to obtain as to ask.

Masonry isn't anyone's work. It's all of ours.

Grand Master Thomas's theme was "TEAM: Together Everyone Achieves More"

TECHNOLOGY AND PRIVACY

t's funny how an edict about taking photos in Lodge can create such a stir. Not recording tiled proceedings outside the Secretary's duties or Master's direction shouldn't be a surprise. But we live in a world of Facebook and Instagram, where indiscretion only takes a fleeting moment.

However, most confusion and debate is regarding things like emailing minutes or having them available on a password-protected web page. Many of us are business people (or even webmasters), and we strive for the amazing transparency and group collaboration technology now provides. It seems a crime not to use it. But we forget one thing:

We are Freemasons.

The nature of our institution is the keeping of confidences, so much that we are labeled a "secret society". Is this an affectation of a past fraught with persecution? Are modes of recognition merely a lingering tradition from a time before dues cards?

Why do we tile and purge the Lodge with great care? It is not that we necessarily do or say much that would shock the public. Privacy bestows freedom of speech – a sacred space to be open and honest. The judgment of the profane world is taken out of the equation, and we cannot be taken out of context by those who do not know our ways.

Like debate within a jury, or sharing within group

counseling, this is an unspoken landmark rather than a convenient rule. Consider that the world's democratic republics were in no small way the product of the Brethren who came before us, and the secret ballot is an immovable part of its foundation. Even discussing one's own vote is a Masonic offense, for without the ability to reveal there is no opportunity to compel.

These are not dependent on social conventions. Technology does not dictate our prudence. It is our prudence that dictates our words and actions irrespective of the times and tools we use.

I'll admit this aspect of Freemasonry is hard for me. I am a very open person, and wear my soul on my sleeve. On a gut level, I don't believe in secrets, at least in the sense of keeping things from people. But that's not what we do. We are not keeping Light from the world or harboring sins.

We are protecting ourselves from misunderstanding, within and without. And persecution is not dead – the families of Masons are still in danger in some parts of the world. And an innocent or lawful word spoken today could be condemning in a darker future. It may seem silly to apply such diligence to the average meeting where there is little more than the paying of bills. But perhaps such a sentiment reveals we are not using our sacred space for more noble and glorious purposes.

We each were given a new name. Let our conscience as to what this means be informed by Constitutional wisdom. Details will be hammered out, but together we will temper our modern practices with our timeless Masonic values.

{Unpublished, December 2015}

LET'S NOT KEEP UNNECESSARY SECRETS

What's the best-kept secret in Masonry these days? When and where Lodges meet.

Let's get "where" out of the way. I've been working on a Lodge locator for some time, and the challenge isn't technological. One of five Lodges don't have a specific street address in the system. My guess is that these Lodges were entered a century before ZIP codes and GPS. Coincidentally, one out of five Lodges don't use MORI, where that information could be easily updated.

The other issue is contact information that a secretary or Lodge may or may not want public. Every Lodge could have a dedicated phone number and email forwarded to a current officer. There are many Lodges a traveling Brother would have no idea how to even contact. Heck, some Lodge websites don't even say what state they are in.

As for "when", brothers know when their Lodge meets, but visiting brothers may not. A calendar isn't just about your Lodge, but other Lodges and bodies that want to schedule events so they don't conflict. So where is the Lodge calendar? In the secretary's hand-written notes? In a bulletin only members see?

The "real" calendar is the one people have the most access to, which today is unavoidably online. Regional Masonic newspapers are fine, but by mailing time, some dates have passed or changed. On the other hand, digital calendars are displayed, shared, combined, and imported into people's phones.

I know brothers who check their district's calendar all the time. When they have a free evening, they see what Lodge is having a communication and visit. It's easier than ever, but like with paper, it's only as good as the commitment of people to maintain them accurately.

I've been working with a number of districts to fix all this. Some are starting from scratch; others have the mechanisms in place but aren't using them well — or at all. On one road trip, I had no way to find out if there were any Lodges meeting that night except to bother a gracious and helpful DDGM.

With today's technology, if people don't have the ability to look up events any time, from anywhere, that is our fault. Yes, we decide for ourselves if we can "get by" without a smartphone or email or social media. At most it's a slight inconvenience to make an extra call or postage stamp to let us know what's going on. But if you're a secretary or in a position that requires communication, you shouldn't decide for everyone else what they can "get by" without.

For example, Brothers' email addresses may not seem like much to people who don't use them, but are a vital piece of contact information. A secretary working to keep them current in MORI is a necessary service to the Craft, as Grand Lodge uses these for communication such as Hiram's Highlights.

Grand Lodge is committed that we, as a TEAM, meet today's demands. Are you in?

TURNING SCREENS INTO BRIDGES

Over the last two years, the Communications Committee has done amazing things. Digitally, we reach more brothers than ever through social media and the Atholl list. Our newsletter, Hiram's Highlights, now brings together the best news from district e-newsletters. And the net of the Internet catches gentlemen wishing to become Brothers in record numbers.

In working to bring substance to the Craft's web presence at the district level, I've made Brothers across the state into friends. I've turned a generous opportunity from Grand Master Thomas and now Grand Master Williamson into some of the most fulfilling work I've ever imagined to do. I've learned that working as a TEAM in Masonry is truly a wonderful Way of Life.

Such work does not reboot every two years. The new chapter was being written before the old was finished. With webmasters in place in almost all districts, we can now connect Lodges and bodies into a functioning web of mutual dependence and support. It's not about websites, but people. People need to know and be able to contact people in the dissemination of information, expectation, and expertise. People need to be accountable while being given the tools they need to work this "craft within the Craft."

From the simple idea of holding a "Webmaster's

Conference" last Saint John's Day, we formed a square club of Brothers interested in using digital media for benefit of the Fraternity. It was standing room only. This year we will be meeting again in Utica, at a larger venue to reflect the progress of the last year. Topics will include mobile-friendly web design, calendar management, and photography. But it's not all technical. We are also planning a presentation of pins and maybe a few surprises.

But wait, there's more! We recognize that technology can improve our lives, and it is not limited to the young. The Masonic Care Community is using digital tablets and music players to Skype relatives and for music therapy, entertainment, and brain-exercise activities. Just as our Lodge work does not end at the last gavel, our digital work does not end when we power down our computers.

Therefore, as an initiative of our growing Digital Square Club, I hereby challenge each and every district to donate at least one digital tablet for use in the Masonic Care Community. I'm working with the MCC's Director of Development, Victoria Cataldo, to determine what devices and hardware will bring the most benefit, and donations can be sent directly to her or brought to St. John's Day. Many more details will have been announced by the time you read this.

Furthermore, I challenge every webmaster to take one day out of their year to visit Utica and work personally with residents to get the most benefit in their quality of life using these tools. Remember, it's never about the tools, but the people using them.

Let's turn little screens into wondrous bridges.

For the full and current details of this initiative, go to {Digital Square Club Link on NYMasons.Org}.

LOOKING BACK, LOOKING FORWARD

As you read this, a cathedral in Barcelona, Spain is under construction to be finished ten years from now, a century after the death of its architect. The plans and tools and techniques to execute the work evolved, surviving generations of social and political conflict and progress.

Our operative Brethren began projects they could not finish, and finished ones started by others. Times and technology change, but the work connects them all, past, present, and future. Such concepts should not be foreign to us.

When I first became webmaster for Grand Lodge, I had lofty ideas about what should be done and how to achieve it. I quickly discovered many of these things were pushed for in the past or already implemented. The raw material was available and the foundation had been laid, albeit under a cover of dust in places. I was not needed to be an inventor or master architect, but a simple, steadfast workman, to continue and further the work. We have willing labor and plentiful tools, needing only organization and effort.

What lesson have I learned?

We should seek out elder Brothers. Even if not up on the times, they preserved the Craft for us. We can change plans to suit today's needs, but let's cherish their work for what it was. But it's more than that. Common sense must prevail, and being a computer science guru doesn't mean you understand how to best tackle real-world challenges. Brothers retired from print media or public relations or marketing have wisdom that applies today, and we need it far more than gadgets and lines of code. They have a right to sit at the table when we think we are being proactive or original. It is our turn to apply their knowledge anew.

What further lesson can we learn?

As those who have gone before us, we cannot grasp the working tools of life forever. We shouldn't take our trestle boards with us. Realize that we don't work for ourselves, because the Craft as a whole belongs to no man. We are stewards of a Way of Life, given by great men whose photos and paintings grace our halls, and hold it in trust for Brothers, some of whom have not yet been born.

Keep a record of what you do and how you do it. As a webmaster, make sure you are not the only one with the keys to a website, Facebook group, or calendar. And always act as if you will pass the torch tomorrow. Can what you have built live on? Can it be curated by someone else, or is it proprietary, created and managed with tools only you have?

And don't be surprised or resistant to new ideas. I am finding Lodges and districts that are ahead of the curve with using technology in ways I haven't thought of.

It's time to cherish the past and embrace the future. They are both bountiful blessings to where we are here and now.

FACEBOOK AND PRIVACY

People my age are grateful the Internet did not exist in the days of our youth. Perhaps we did things that shouldn't be on YouTube or said things we wouldn't want on Twitter or shared on Facebook. But today we live in a world where permanent indiscretion only takes a fleeting moment. Even if we don't say anything we regret, there's the issue of privacy. This is why some of us want nothing to do with social media.

But can we ignore it? If your Lodge doesn't have a site, you missed the chance to be cutting edge by 20 years. That boat has sailed. A website can still be very useful, but the main form of communication is now social media. Your members don't use it? That's fine, but nearly every person considering the Craft does, several times per day. And there are multitudes of Brothers across the state (and world) of all ages that actively use it. It's the primary place Masons discover – and then attend – events across your district and beyond.

What is the result of letting that boat sail? An "old" Lodge will not be a fit for "new" members. We are confronted with the choice of embracing and witnessing the Masonry of a new generation or waiting for the last Brother left to turn in the charter. The challenge is for all generations to meet on the level, and without being in touch with today's ways (as well as the old) that isn't possible.

But even if you are engaging each other and the public online, there are very important things to remember. First, what you post and the way you communicate is a reflection upon the Craft. Even between brothers, it's all too easy to disrespect another's opinion from a keyboard in the middle of the night. Do we live up to the virtue of Civility promoted by our previous and current Grand Masters? Online or off, we are charged to be upright.

Secondly, just because the tools are there doesn't make them the right tools for all our labours. Many of us are business people (or even webmasters), and we strive for the transparency and group collaboration technology now provides. It seems a crime not to use it.

But don't forget we are Freemasons.

We do not tile and purge the Lodge because what we do or say would shock the public. Privacy bestows freedom of speech – a sacred space to be open and honest. The judgment of the profane world is taken out of the equation, and we cannot be taken out of context by those who do not know our ways. That is one reason minutes should not be available in digital form, anywhere. The prohibition still applies for photographs or ritual. What we experience should impress the mind, not the screen.

Technology does not dictate our prudence. Our prudence dictates our words and actions, regardless of technology. Whatever tools you use, use them with well – and with wisdom.

A TRAVELING MAN

I've taken the notion of "traveling man" to heart. In my few years as a Mason, I have seen and done things I would not have imagined if I had stayed within the walls of my Lodge. Therefore I did not shy from setting up a webmaster and communication conference for the metro region. And I have received a plenty of a master's wages for it.

I treated it as a journey instead of a quick trip, and so it was. On the drive down, I spent a day at the Masonic Care Community. I challenged webmasters to volunteer and it was about time I did it myself. I saw firsthand residents putting to use the digital items we had collected. A resident was using Skype to see her daughter across the country, while a group of residents played Linked Senior word games around a large touch screen.

The programs, resources, and quality of care cannot be adequately described in print, or in presentation at Lodge. Like Freemasonry itself, it cannot be appreciated except by experience. The care community is the "Jewel" of our Fraternity, but I see the whole campus as the "Soul" of our Grand Lodge, the embodiment of our obligations, values, and hope for the future. Visiting, even once, will expand your perspective and pride as a New York Mason.

Later that night I used the Lodge locator on NYMasons.Org

and had a wonderful fellowship with Remsen Lodge No.677 as well as a tour of the Utica Temple.

The conference itself was everything I had wanted it to be. Instead of preaching what needed to be done, I called on local Brothers to share their knowledge, experiences, and challenges in serving the Craft with websites, calendars, social media, email newsletters, and even phone apps. One lesson was clear: we have the talent we need within our local quarries. Questions were asked. Ideas were shared. Relationships were built. We had Brothers from across the metro districts — webmasters, DDGMs, Grand Lodge committee chairmen, and even guests from the MW Prince Hall Grand Lodge of the State of New York and the Grand Lodge of the District of Columbia. The presentation space on the second floor was impressive and a local Brother catered the event. And I finally got to tour most of the Lodge rooms.

My world had become larger, only to find it small again. At a gas station in Poughkeepsie, I ran into a Brother from my own district who had just attended a Grotto event.

This was one of my most memorable Masonic travels, but I know my journey is just beginning. If you obtained the Master's Word and have not traveled, it is like an uncashed lottery ticket. When you visit Utica, or Grand Lodge, or Camp Turk (still on my list), you truly experience how far our Brotherhood and its work extends, and how near we really are to each other. Travel makes Masonry a Way of Life. Safe travels, my Brothers.

LEARNING NEW TRICKS

W e are a society of ideas.

Every business, every club, every charity or group, gives lip service toward looking to the future and wanting "fresh blood" and new ideas. However, accepting new ideas is easier said than done. Gaining new employees or members is more about numbers than revitalization. The idea of catching up with the times is a vague platitude. It can be delegated to "younger" people hoping for magical results, but those in charge don't want to know details. Those assigned can "fix it" so long as they don't impose any change to the way the rest of us have to do or see things. You can imagine the frustration and result.

Many organizations hire other companies or people to do these things. They let themselves go for so long, they don't know how to promote themselves or conduct business in a modern world. What has always worked for them doesn't cut it for everyone else, but they may not see it or just don't care. others sense something is wrong and want it fixed — so long as they don't have to learn any new tricks.

Surely Freemasonry is another story. We are a fraternity of ideals.

There is no doubt in my mind about our ideals. But how consistently are higher thoughts coupled with nobler deeds? The Craft's allegories for a productive life is physical labor.

But whereas our operative forefathers didn't merely draw plans and order their apprentices to move stones around, we assign committees and vote on motions as if it is the real purpose of our work.

I could make this about technology, but it's deeper than that. We don't question traditions or ritual, but question every tiny change to it. For even the most "proficient" Lodges, degree work is sometimes akin to just moving stones around. We keep "busy" – making committees, assigning a webmaster, "looking into" new ways to communicate or appeal to newer generations and keep them interested. But how often do good intentions result in finished work we can judge to be good and true?

We take for granted the strength of our institution, our Lodges, our ideals. We stopped guarding the door to habit and complacency. We say we want things to be better but we don't want them to be different. We can't have it both ways.

We want new Brothers to bring freshness to a dusty institution. Or are they little more than dues payments and manpower to continue doing things the way we always have?

Not everyone can do every kind of work, especially with technology. But as the Master is responsible for all that occurs in the Lodge, each of us has a responsibility to not be ignorant or resistant to efforts to modernize. It can't be the new guy's pet project. As we should strive to do more than move stones around or memorize ritual, know that all our work will be judged for its result – and how it adapts to here and now.

AS ABOVE, SO BELOW

"A s Above, so Below."
Who would have thought this metaphysical truth would apply to the Internet? The Web is an unfiltered reflection of the good, the bad, and the ugly of humanity. Every aspect of who we are, shining light and dark corners, can be found online. The secret is we can use it to work the other way.

My first major client was an organization of four interconnected entities. Sitting at their board meeting, no one seemed to know who was in charge. Everyone had their own agenda and view of what the overall organization's focus was. I was there to make sense of it to build their website, but I discovered that to do so I had to give them something by which they could make sense of themselves. Their website became an organizational chart of sorts. How they explained themselves to the public gave them a palpable understanding of their own respective roles and structure.

What did I learn? An organization's website reflects its identity — its message, its purpose, what it aspires to do and to be. It also can reveal if an administration is confused or fractured about its priorities. Too often a complex organization's website can become a circus of competing ideas vying for the visitor's attention. (This is an understandable ongoing challenge for NYMasons.Org, as Grand Lodge is a vast diversity of committees, programs, and projects.)

This is why Lodge and District leadership should be directly involved in their web presence. The very process of building and maintaining it forces us to focus. We are setting a virtual trestle board we must live up to, one that can't be hidden in a committee or tucked behind a secretary's desk.

This applies to us individually as well, and not just online. Does our bumper sticker reflect the courteousness of our driving? Does our lapel pin reflect the way we do business? Does our Masonic profile image reflect our language in social media? This is all the same question: Do we live up to what we purport to be? And do we realize the power it gives us to represent ourselves for all to see and then live up to it?

Even our buildings follow this two-way rule. Does the sacredness of our Work match the beauty of our buildings? Our Lodge Rooms should inspire, not detract from our experience, and if we do not care for them, what does that say about us? What care do we put into our virtual buildings – our websites?

Is a website just another afterthought of who we are? Or will we harness the process of making and developing it as a tool to (re)define who we truly want to be? Will it be another copy-and-paste platitude about our history and legend? Or will we be the same in public as behind tyled doors?

Let our web presence be a trestle board. Let us say what we will do, loudly and with zeal, and then endeavor to do it.

TO BE YOUNG AGAIN

Want to be young again? One District Deputy suggests the secret is to be around youth. After all, we tend to rise to the demands and expectations of those around us. How does our Fraternity find itself young again? Answering the door is a good start. Being welcoming is better.

I hear much debate about Millennials and what to expect from them. I choose to ignore such criticisms or praises. Why? Because we take our obligations as individuals, not classes, races, religions, or generations. Each has their own path in life, and in Freemasonry. We disrespect them by imposing assumptions and risk missing out on a valuable member of our Fraternity. Even after by-laws are signed, we may label them this or that, depriving them of their individual character as one of "them" instead of one of "us".

Let's ask the question again: Do we want to be young again? Then don't ignore new members. Being a Brother to the "new kid" is not optional unless you don't want them back. No one should ever sit alone in Lodge. No one willing to work should be overlooked, and none should go away dissatisfied, unable to reap our wages.

We were all new Brothers once. Rob Morris was a Mason only three years when he wrote the first Eastern Star ritual. Some Masters of Lodges are under 30 years of age. Someone we raise from this generation will be Grand Master one day and bring with them the life experience of their times. It will be their Freemasonry no less or more than we can pretend we have claim to Freemasonry today. No one owns it. It was passed to us and will be passed to them, and even the most stubborn cannot stop the Level of time from making it otherwise.

We see in the world fast-paced changes in technology, social expression, cultural exchanges and conflicts. We think today is so different from yesterday and fear for our continued existence. But the Craft has always fared such things. The more things change, the more we find human nature does not. Masonic values don't change, nor the world's need for them.

Future generations will not preserve Freemasonry by repeating only what we did and how we did it. Every "that isn't the way we've always done it" is a nail in our coffin, and we must know that because it's a running joke in many Lodges. Each generation must pull out those nails and make our Landmarks and Rituals their own, interpreting and translating it in ways that make sense for every age to come.

Our Gentle Craft will be young again. It is already happening if only we have eyes to see it. There are many ways to use technology for our benefit, while still providing an alternative to the superficialities of digital living. Underneath it all, it will always be about how we treat our new Brothers, teaching them what we can, and supporting them when they gavel becomes theirs.

"WEBMASTER" ISN'T WHAT IT USED TO BE

I t's already too late. Over the last couple years, I've been pinning lapels across the state with the "GLNY Webmaster" pin (silver for Lodge and gold for District Webmasters). But the term "webmaster" is not what it used to be. It may as well be "computer guy." Not just because managing a website is more of an incidental task rather than a project these days, but there is so much else to do online – social media, e-newsletters, online calendars, and who-knows-what's-next.

When I was tasked to ensure each district has a website (a goal ever elusive), I discovered a need for competency in more than web code. At every level, we face difficulties implementing technology that has already passed us by. We're mailing CDs and printing emails as handouts. People type in calendars by hand for websites that no one can share or subscribe to on their phones. There's so much more that needs to be managed than a few years ago.

Now and then we ask every district or Lodge to have a PR person, or communications officer, or webmaster. We need all of them, and they are all related. It could be a team or a person, but its too often the DDGM, or no one.

It's becoming archaic, but I use "webmaster" to describe someone who takes responsibility to be aware of and advocate using technology in the Craft. It really doesn't mean updating an officer's list once a year online. It doesn't mean one person doing all those things above. The secretary handles communication, but that means more and more ways every year. There must be a go-to guy, someone to coordinate the Secretary, the Brother who posts on the Facebook page, the Brother who sends out an email newsletter, and yes, the Brother who manages the website if there is one. He could be one or more of those people. He could be the communications officer or PR chairman, but in the end someone has to actively be this point man.

This is who should get the lapel pin. I should have a name of someone I can call in every district; this person should have a contact for each of the local Lodges and bodies. It's that simple. If we start with Brothers instead of getting hung up on titles and job descriptions, we can all work together to fill in the gaps as we are able.

The initiative that led to building this network didn't go away with our new Grand Master. The work is always there, especially with new districts. Ask your Lodge and District leadership who is your webmaster. Volunteer or recommend someone. Join the Digital Square Club to meet other Brothers who want to serve the craft digitally. Reach out to me for resources; keep me informed for statewide communication.

And until someone comes up with a better name, keep calling me Grand Lodge Webmaster. It's not a real title, but it's real work I can't do without your help.

KEEPING IN TOUCH

I am amazed how grateful some Brothers are to hear from someone. It seems many Masons don't hear from their Lodge apart from occasional newsletters and dues notices. Why is it so hard to keep in touch? There are many reasons.

We don't live in small, tight communities anymore. Our Brothers were our neighbors. Lodge notices and news were in the town paper. Today, members have to travel and don't get the "local paper". Masonic papers are few and far between.

The 9-to-5 workweek is no longer the norm. Forget television and family as excuses for Brothers not being there. Work schedules vary greatly, and change often. Getting more than two people together in person for a committee can be a tedious chore. On top of this, our aging population means some Brothers won't drive at night or are even infirmed. The bottom line is that much of our membership cannot attend regularly and are out of the loop.

Also, there is no single medium to reach members. Some have email; some do not. Some use Facebook; some don't. Some things require a phonecall, if we even have their (correct) number. Some things can be mailed, but that means time and expense. It's a nightmare job for the secretary and others to hope to reach everyone about everything.

Technological changes in travel and communication have affected our way of life and brought us these challenges – but not without also providing solutions.

But first, we need to stop thinking of Lodge as something that only happens in the Lodge room, and that leadership and participation depends solely on attendance. (Most tasks can be done in committee, if for no other reason than to avoid business displacing Masonic programs and education.) Lodge is nothing but us and our relationships with each other – and that means communication more than just sitting together in the same time and place. Counting on announcements made in Lodge feeds the attitude that if you want to know, you would have been there.

We can't realistically reach everyone about everything all the time. We must plan multiple means and how often we use them and for what. Newsletters are solid for general news and information. Social media is great for casual sharing. Email is good for planning. Robo-calls are best for urgent matters. For committees, there is phone and video conferencing. But we have to get out of our comfort zone to use these things.

All of us can do our part. Make sure the secretary has your correct contact information. Your email address – if you use it – will save stamps and you will be more promptly informed. Be in touch as much as you want through social media.

Times have changed. Technology has brought us challenges – and solutions, if we are willing to use them. We are the stones that comprise our living Temple, the material and purpose of our Work. And our relationships are only as strong as how well we keep in touch.

TEAMWORK MEANS SHARING — FOR ALL OF US

---

eamwork. How do you take a diverse group of volunteers and their ideas and develop a comprehensive plan? How do you execute that plan with so many gaps in the buy-in from others? I suppose that is what leadership is for. I don't claim to have mastered such a thing. I just know I have a long way to go, and that many others are in the same boat.

But I know what we CAN do. We can communicate better. We can share information to make informed decisions to build consensus.

Effective sharing – the dissemination of knowledge and resources — may require particular individuals to set up places to store information. This can be anything from a public website or social media channel to a private cloud of folders and files available on a need-to-know basis. Some are still struggling to find out where to get paper forms or directories, not knowing they can be downloaded or emailed to them. This sounds easy enough for webmasters and the technology people in Grand Lodge, but it is not for one reason: We "techie" types work in a bubble others have put us in.

I see this in my professional web development experience as well. People want to "fire and forget" anything to do with technology. They subconsciously outsource their own responsibility regarding such things on the grounds they don't understand how it works. It must be "our" job. I'm sure at one time people were afraid to drive cars because they didn't understand the combustion engine. It's natural to avoid what we don't understand. But it handicaps or even prevents us from doing what we need to accomplish.

This brings us to the other half of the equation, and a solution if we take it. Communication is everybody's responsibility. It is the leadership, not the people who handle the infrastructure, who are in a position to disseminate and educate about what is available. Every Brother can make a note and share what is going on at every level of the Craft. Webmasters can work to provide the means, but are not the ones who should be driving the car. Someone must take the wheel and others must be invited to ride.

Every Lodge and committee would see an immediate benefit if we learned to share information in a way that doesn't become background noise. There are Masons who have been in the Fraternity for decades who don't know about things published in every edition of this magazine and promoted every year by their DDGM. I don't have a technology solution for that. I only have a human one, and that is we must all lead each other by the hand when we discover a gap in our knowledge. We must know where to find the information and how to share it efficiently.

To best work and best agree, we all must bridge the gap between those who manage the information and those who need it.

LIKE A BUSINESS, NOT A BUSINESS

I would hope that who we are as Masons affects who we are in our business dealings. But what about the other way around? What can we learn from the business world to improve the Craft?

Knowing our place in the market (role in society) is vital. Knowing our brand (how people see us and we want to be seen) is necessary. We need quality control to ensure standards (landmarks and values). We need to measure our success (retention, participation). But most of all, we need to deliver on what we promise, not just fine-tune our marketing and increase sales (raise Masons).

Here's the problem — we can take these analogies too far. What is our product, really? We do service, but we are not "a service". The success of the Craft is personal. We are both the consumer and the product — each of us a Living Stone in a mutual constant improvement process. The real Secrets of Freemasonry cannot be bought or sold.

How do we reconcile those ways we need to act more like a business and ways things we clearly are not? Those aspects that lend itself to corporate culture are administrative, or as I like to describe it, housekeeping. It needs to be done as we live in the World as much as we are not "of" it. Grand Lodge is working hard to reach potential candidates, to be more visible online and elsewhere. They can ensure quality control through the accountability of Masonic Law and develop more effective education.

But the real work of the Craft is in each Faithful Breast, and the individual Lodge is at the heart of any jurisdiction. Grand Lodge doesn't make Masons. No one joins Grand Lodge, or some District therein. They petition and become a member of a Lodge.

But there's a challenge. We live in a world where we have gotten used to "those people" on top being in charge and responsible, rather than appreciate they are servants we have selected among ourselves to manage larger issues. It is up to us in the trenches (quarries) to meet the need. Grand Lodge can set a standard, but we must meet or exceed it. The Lodge is not a store, or a website. We as Brothers are the only real "point of sale" in our communities. Our lives are the testimony and recommendation for our "brand".

Grand Lodge is now producing our own videos extolling Freemasonry. We're stepping up our game in social media. We're getting in step with our times in terms of technology. But what happens next is ultimately in the Lodge's hands. There is nothing worse than selling someone on something, even an idea, when it turns out to be less than advertised. We cannot look afar for blame when we raise Brothers who don't come back after their degrees. We cannot put our best foot forward only on the business end of things, but work best where it counts the most — in our hearts.

GRATIFICATION

G ratification has become a bad word. It taken more as indulgence than satisfaction. But in today's fiber-optic and wireless society, we only hear of "instant gratification" — the opposite of reward for hard-earned work. How do we compete with this? I say we don't have to.

Gratification isn't far from the idea of Masonic wages. We get our corn, wine, and oil, in outward forms such as anniversary pins, special aprons, finished reading courses, and other certificates and awards. We seek milestones – hopefully not only for their own sake – but to see each solid step on the Winding Staircase to remind us of how far we've climbed and how far we have to go. It is the reason many belong to concordant bodies — additional Degrees continue the gratification we experienced being initiated, passed, and raised. We could seek and find still further Light without them, but they give form and structure to such achievement.

Surprisingly, we can learn from those technologies we eagerly blame on the decline of fraternalism. Many game systems use artificial learning to adjust rewards so that they don't come so easy to make the player feel they didn't earn it, but not hard enough for them to stop playing. That's why they are so addictive.

Likewise, Brothers may be intimidated by the memorization of lengthy ritual, but they don't want to get a pin for just paying their dues, either. They expect to be challenged and rewarded, continually. There must be something in between Degrees and anniversaries to keep them in the game. There must be work for every Brother that is meaningful and accessible, and visible not only to the community but to themselves.

Sadly, many meetings are without reward. I'm not saying we should get a sticker for showing up. We could have traveling gavels for individuals, or passports to sign upon visiting district Lodges and beyond. We must have some palpable token that we got what we came for. Any program, or even a simple discussion can challenge us to answer the question, "Am I better or more enlightened for being here tonight?" Consciously answering this question is its own reward, and will tell us if we are living up to our obligations as Masons and Masters of Lodges.

What about ritual proficiency? Some districts and jurisdictions have a certification or card for those who are deemed able to do ritual to a measurable, high standard. There has been some talk of creating a database of such individuals that Lodges can use to find "ringers" for certain parts, but even if there was not a need, it would be an opportunity for personal pride in our Work.

There is an ideal to do our best without seeking any prize. But our ritual is clear there is a connection between work and reward. Perhaps this is a recognition of the human condition, and we cannot feed our souls on intentions alone, but on all the Wages of a Master Mason.

OUR SPIRITUAL TECHNOLOGY

always use the term 'technology' broadly. Our Working Tools are derived from ancient technology. Websites and cell phones are modern technology. But whatever age, technology is useful insomuch as it is used, and used wisely. But the constant throughout history is human nature. We have the same basic needs as Adam or Noah or Solomon. We want to build, learn, communicate and connect. Some tools are better than others, and depends greatly on circumstances.

A perfect example is our ritual. Sure, anyone can look up some of the things we keep secret on the Internet. We have ever-improving printed versions, and someday may even be a password-protected phone app. But all that is beside the point. We are asked to be on guard against those who appear to know the work but are not trained properly in our Art. It's about transmission of meaning, not merely a set order of words. Those of us who understand this agree that the secrets we possess are truly mouth-to-ear. They can't be lost in fire or flood, or stolen from a briefcase, or bought and sold or given away online. Our real technology is spiritual, carefully preserved in those unseen places of the soul.

As an institution, we may be supported by plenty of physical buildings, books, and gadgets. We may have social media accounts and email blasts and robocalls. But when we settle down to the real business of the Fraternity, it's about a grip of the hand, gentle council, a real phone call or visit in person. This does not mean we should not use the world's technology. It just means it must always take a back seat to the best practices of Moral Science, which are ultimately human, personal, and intimate.

Today, we can reach nearly everyone any time by so many channels. That's "broadcasting" and impersonal. When was the last time the leaders of a Lodge reached out talked to every Brother individually? Can we agree doing this yearly (or more) is better than just using online people-finders when numbers go bad and finding obituaries through search engines? Do we visit Brothers who can't make it to Lodge? Do we provide rides? We have the technology of the phone and the automobile but in a digital age even these things seem like extra work. But isn't it the most important work?

Perhaps to preserve our timeless traditions and values, we need to not just use the technology that's available to us, but go back to the "old ways" where it counts. What is most efficient is not the same as what is most meaningful or right. Even in our collective dealings, we must ultimately aim for each Point within a Circle. The Craft's existence makes the world a better place, but is different from all others because it teaches us to apply allegory and symbols of morality to better ourselves and each other. And there is no substitute for sincere Points of Fellowship to do just that.

MY JOURNEY -
ACKNOWLEDGMENTS

Even two years after being Raised to the Sublime Degree, almost none of my Lodge Brothers knew what I did for a living. I petitioned the Masonic Fraternity to be among those who shared my values, with a particular interest in joining the Western New York Lodge of Research No.9007 F&AM, and the not at all insignificant purpose of my wife fulfilling her lifelong desire to belong to the Order of the Eastern Star.

In 2013, a local Brother, Jeffrey Williamson, called a town hall meeting for Erie County's Masonic districts. There was much talk about communication, calendar conflicts, and the lack of a real web presence. I don't know what got into me, but I stood up and announced that I owned a web development company. What followed swiftly from this opening my mouth is almost a blur. I never would have guessed my usual vocation would intersect so directly with my journey in Freemasonry.

It was a while later I got the go-ahead from one of the District Deputies to set up a site, but by that time, Right Worshipful Williamson (then a Grand Warden and Past Grand High Priest) connected me with the upcoming Grand High Priest of the Grand Chapter of Royal Arch in New York, Piers

Vaughan. I wasn't even a Royal Arch Mason, and here I was, setting up a state-level website of a concordant body.

Sometime early in 2014, RW Williamson (who had become Deputy Grand Master) and the new Grand Master, Bill Thomas (also a Past Grand High Priest), discussed candidates to take over the webmaster duties of NYMasons.Org, the official website for the Grand Lodge of the State of New York. They both knew my work, and so it began.

This wasn't gratis work — I did get paid a reasonable sum for my web development work and cover my cost of hosting and such. What I didn't realize (and discovered by accident) was that I had been appointed to the Communications Committee, and therefore became a voting member of Grand Lodge. Being green in the Fraternity, I didn't understand what this meant, but from that point on I endeavored to define the needs of the Craft peripheral to my work, and devote much of my own time to meet them. In the end, the website itself proved to be a small part of my overall work for Grand Lodge.

The greatest need was digital advocacy and education. Deputy Grand Master Williamson and his team traveled throughout New York State, holding town hall meetings to educate Brothers in the quarries about Grand Lodge, its activities and initiatives, and to listen to their needs. I tagged along whenever I could, even riding in the "Grand Van", feeling a bit like a hitchhiker among nobility. At the gatherings, I spoke about the need for Lodges to use technology in their communications and how they represent themselves to the public.

From talking to Brothers across the state, I realized there really was no cohesive, large-scale coordination or education regarding communications other than precedents and mechanisms set up in the times of the Pony Express. Not really having a job description, I took it upon myself to work toward this, wrote articles, held regional conferences, and invented the "Digital Square Club," all with the support and guidance of Grand Masters Thomas and Williamson.

I had somewhat backed away from these appended duties while focusing my energies into Ken-Ton Lodge No.1186 as a first-time Master. But in virtue of my central function as webmaster of NYMasons.Org, I continued to work with brothers across the state in various committees and projects.

I am not sure what my future role will be. There is still much to be done, and I am so grateful for the benefits of meaningful work and new friendships. But even as my usual vocation, I can't do this forever. It becomes more difficult each time I must reinvent myself in terms of professional services and skills, and time waits for no one. Having started a publishing company (the one producing this book) and slowly transitioning my life goals to align with that, I am in the position to share what I've learned and hopefully inspire a new generation to take on these and future challenges.

I commend in advance whoever they may be. But here and now, I raise my glass to those who have gone before me. To this end, there are many with foresight and have worked toward the future, but I want to personally acknowledge RW Ronald Steiner and RW Ed Chiani, Brothers who have taken steps and plotted the course of Grand Lodge in its starts and restarts into digital modernization. Without their work, I would have had no foundation.

It is with deep gratitude that MW Bill Thomas and MW

Jeffrey Williamson had faith to entrust me with work that opened a larger world of Freemasonry to me. There are so many friendships and connections that would have been unlikely or impossible if it were not for the experience of this opportunity. Their time in the Grand East completed, their Masonic mentorship endures, which is even finer gold.

I have so many mentors, too many to list for fear of forgetting someone. But there is one person who is most responsible for my involvement in Freemasonry — my wife. She isn't merely the tolerant wife wondering when I'll be home. She loves being a part of the Masonic Family and has encouraged my involvement even before I knew what Masonry was. And there is little I can even pretend to have achieved without her.

www.ingramcontent.com/pod-product-compliance
Lightning Source LLC
Chambersburg PA
CBHW071646210326
41597CB00017B/2131